TUNGSTEN CARBIDE – PROCESSING AND APPLICATIONS

Edited by **Kui Liu**

Tungsten Carbide – Processing and Applications
http://dx.doi.org/10.5772/3148
Edited by Kui Liu

Contributors

I. Borovinskaya, T. Ignatieva, V. Vershinnikov, A.K. Nanda Kumar, Kazuya Kurokawa, Marcin Madej, Zbigniew Pędzich, Paweł Twardowski, Szymon Wojciechowski, Yufeng Fan

Published by InTech

Janeza Trdine 9, 51000 Rijeka, Croatia

Copyright © 2012 InTech

Notice

Publishing Process Manager Sandra Bakic
Technical Editor InTech Prepress, Novi Sad
Cover InTech Design team

Additional hard copies can be obtained from orders@intechopen.com

Tungsten Carbide – Processing and Applications, Edited by Kui Liu
p. cm.
ISBN 978-953-51-0902-0

For EU product safety concerns:
IN TECH d.o.o., Prolaz Marije Krucifikse Kozulić 3, 51000 Rijeka, Croatia,
info@intechopen.com or visit our website at intechopen.com.

Meet the editor

Dr. Kui Liu received his BEng and MEng degrees from Nanjing University of Aeronautics and Astronautics (NUAA), China in 1989 and 1996 respectively, and PhD degree from National University of Singapore (NUS) in 2002. He worked as a lecturer at NUAA from 1989 to 1998 and was a postdoctoral fellow at NUS from 2002 to 2004. Currently he is a Team Leader and Scientist at Singapore Institute of Manufacturing Technology. His research interests include ultra-precision machining, nano/micro machining, ductile machining of brittle materials, ultrasonic vibration assisted machining, high speed machining and tool development. He has authored and co-authored more than 100 technical papers published in international journals and conferences. He is a member of SME, euspen and JSPE.

Contents

Preface

Tungsten carbide (WC) was first extracted from steel and properly identified around mid 19th century. It has attracted great interest to both engineers and academics for the sake of its excellent properties such as hard and wear-resistance, high melting point and chemically inert. Although it has been known for over one hundred years, recently tungsten carbide has been applied in numerous important industries including aerospace, oil and gas, automotive, semiconductor and marine, which also has a promising future. Cemented tungsten carbide, often simply called carbide, and also called cemented carbide and hard-metal, is a metal matrix composites (MMCs) where tungsten carbide particles are the aggregate and metallic cobalt serves as the matrix. It has excellent physicochemical properties, particularly enables to resist high temperatures and is extremely hard, which bring out wide application in the industry for cutting and mining tools, moulds and dies, and wear parts.

This book aims to provide fundamental and practical information of tungsten carbide from powder processing to machining technologies for industry to explore more potential applications. Chapter 1 introduces the self-propagating high-temperature synthesis (SHS) method to produce nanosized tungsten carbide powder. Chapter 2 explores the kinetic mechanism for spark plasma sintering (SPS) of tungsten carbide nanosized powder to produce cemented carbide. Chapters 3 and 4 are dedicated to production of metal/ceramic matrix composites with enhanced mechanical properties using tungsten carbide particle as a reinforcement phase. Chapter 5 is dedicated to the machinability investigation of cemented tungsten carbide, which could expand their application areas by making components using novel machining technologies. The last chapter presents an ultrasonic vibration shoe centerless grinding technology for tungsten carbide component manufacturing.

The book can serve as an informative reference for academics, researchers, engineers and professional that are related to tungsten carbide processing and applications.

The editor would like to thank InTech for this opportunity and their enthusiastic and professional support. Finally, I sincerely thank all the authors for their contributions to this book.

Dr. Kui Liu
Singapore Institute of Manufacturing Technology,
Singapore

Chapter 1

Self-Propagating High-Temperature Synthesis of Ultrafine Tungsten Carbide Powders

I. Borovinskaya, T. Ignatieva and V. Vershinnikov

Additional information is available at the end of the chapter

http://dx.doi.org/10.5772/51303

1. Introduction

Transition metal carbides, particularly tungsten carbide, are rather attractive due to their physical and mechanical properties [1]. They are characterized by the high melting point, unusual hardness, low friction coefficient, chemical inertness, oxidation resistance, and excellent electric conductivity. Nowadays, highly dispersed tungsten carbide powders appear to be very important for production of wear-resistant parts, cutters, non-iron alloys, etc.

It is well known, that fine-grained alloys demonstrate better mechanical properties in comparison with coarser alloys of the same composition under the same terms [2-4]. Use of ultrafine or nanosized powders is one of the most efficient ways to produce new materials with required properties.

That is why nowadays the production technologies of nanopowders play the leading role among the widely used directions.

There are several phases of tungsten carbide; the most important ones are WC and W_2C [5]. Though W_2C is unstable at T=1300°C, in most cases the mixture of WC and W_2C is observed in the synthesis products. Precipitation of the single phase of WC is only possible in the narrow area of the technological parameters [6].

There are different ways to obtain tungsten carbide powders, and each process changes the characteristics of the forming product.

Tungsten carbide powders are obtained by direct carbonization of tungsten powder. This process implies production of pure highly dispersed powder of metal tungsten within the first stage. The initial material in this case is very pure WO_3, tungsten acid or ammonium tungstate [7-9].

The second stage includes carbonization of tungsten by carbon in the graphite furnace with hydrogen atmosphere. Depending on the type of the furnace, atmosphere, and carbon content the reaction occurs according to the scheme:

$$2W + C \rightarrow W_2C$$

or

$$W + C \rightarrow WC.$$

The obtained tungsten carbide powder has particles of the indefinite melted form, minimum 3 – 5 μm in size and contains 5 % of W_2C minimum. The reduction terms greatly influence the characteristics of the metal powder and forming carbide.

Thermochemical synthesis of nano-phased tungsten carbide powders was also studied. It consisted of two stages [10, 11]. At first, nano-phased powders of metal tungsten were synthesized by reduction of various tungsten salts and chemical decomposition of vapor of volatile tungsten compounds. Then nano-phased tungsten carbide with the particle size of ~30 nm was obtained by carbonization at low temperature in the medium of controlled active carbon-containing gas phase.

The method suitable for tungsten carbide synthesis at low temperatures (~800°C) during 2 hours was suggested [12]. It is based on the gas-solid reaction between a tungsten source (ammonium paratungstate or tungsten oxide) and carbon-containing gas phase which includes a mixture of H_2 and CH_4.

The conventional calcination–reduction–carburization (CRC) process offers the potential to manufacture commercial tungsten carbide powders with median grain sizes below 0.5 μm (ultrafine grades) [13].

In [14] point to that transferred arc thermal plasma method is more economical and less energy intensive than the conventional arc method and results in a fused carbide powder with higher hardness. Coatings of high wear resistance can be produced using fused tungsten carbide powder with WC and W_2C phases, which can be economically synthesized by thermal plasma transferred arc method [14].

However, it is not economically efficient to use very pure and fine tungsten powder obtained from tungsten compounds at the stage of its reduction for producing a large quantity of tungsten carbide powder.

The existing economical and technological restrictions make the problem of the development of large-scaled cheap production of ultrafine and nanosized tungsten carbide powders very actual. Nowadays, a promising ecologically safe method, discovered in 1967 by academician A.G Merzhanov and his co-workers I.P. Borovinskaya and V.M. Shkiro – Self-propagating High-temperature Method (SHS) – is used for obtaining refractory compounds of high quality. This method combines a simple technology with low power consumption and allows obtaining products with regulated chemical and phase composition and dispersion degrees. Therefore the possibility of application of SHS technology for preparing ultrafine and nanosized tungsten carbide powders represented practical interest.

2. Experimental

2.1. Self-propagating high-temperature synthesis (SHS)

The new scientific direction SHS was developed at the interface of three scientific fields: combustion, high-temperature inorganic chemistry and materials science. SHS is an autowave process analogous to propagation of the combustion wave with the chemical reaction being localized in the combustion zone propagating spontaneously along the chemically active medium [15, 16]. The essence of the process is occurrence of exothermic reactions at temperatures developing as a result of self-heating of the substance; the synthesis temperature is up to 4000°C, the temperature growth rate – 10^3-10^6 K/s, the combustion velocity – 0.1-10 cm/s.

Thorough fundamental investigations of the SHS process have proved that chemical transformation in combustion waves and product structure formation occur simultaneously with high velocity and at significant temperature gradients. These peculiarities of the process provide practically complete chemical transformation of the mixture and a specific structure of the combustion products. Application of SHS allows avoiding the main disadvantages of conventional technological processes – high power consumption, complicated equipment, low product output.

The extreme terms which are characteristic of SHS of chemical compounds affect chemical and phase composition of the products as well as their morphology and particle size [17, 18]. The experiments in product quenching by special cooling methods immediately after the combustion front propagation have proved that "primary" product particles of 0.1-0.2 μm in size can be formed in the combustion front [19, 20].

The product structure formation during the chemical reaction was called primary structure formation while the structure formed in this case was called the primary structure of the product. The characteristic time of the chemical reaction is 10^{-3}-10^{-1} s; the time of the primary structure formation being the same. After the chemical reaction the particle size increases as a result of the secondary structure formation process followed by collecting recrystallization [21]. The duration of the process depends on the sample cooling mode and is usually about some or tens seconds.

Transformation of initial reagents to final SHS products is a complicated multiparametric process. There are various ways to govern it. The main types of the occurring processes are solid-flame combustion in the solid-solid system (one of the varieties is combustion with the intermediate melted layer), gas-phase SHS (chain flames, combustion of condensed systems with gaseous intermediate zone), combustion of solid-gas systems (filtration combustion, combustion of gaseous suspensions) [22].

Let us consider the possibilities of these processes.

In order to obtain ultrafine and nanosized products in the processes of solid-flame combustion, one must use the reagents of the same dispersion. In solid-phase systems with the intermediate melted layer the possibility of nano-crystal formation depends on crystallization and recrystallization processes, combustion heat modes and product cooling after the reaction.

In the case of gas-phase SHS (gas combustion followed by a condensed product formation) the product elemental particles consolidate with each other and form nuclei on the surface of which the following reactions occur. If fast artificial cooling is used, it is possible to arrest the particle size growth at a required stage and obtain nanopowders by depositing the particles from the gas mixture.

At gas-phase combustion the initial reagents, intermediate and final compounds remain in the condensed state (either liquid or solid) during the entire reaction [16, 23].

The SHS method has provided the possibility of obtaining a great number of compounds in the dispersed state (powder). Among the materials for which the technological backgrounds are well developed the main ones are powders of refractory compounds. They are widely used in industry due to their outstanding properties such as hardness, thermal stability, abrasive wear and resistance.

There are several directions of the SHS technologies. The widest and well-developed type of SHS reactions is the synthesis reactions of refractory compounds from elements. It is oxygen-free combustion. Both powders and gaseous elements take part in the chemical reactions. Besides, some regulating additions R are introduced into the initial mixture. They can be synthesis products (as diluents), various inorganic and organic compounds.

Another direction is combination of SHS with thermal reduction (SHS with a reducing stage) when the compounds of elements (oxides, halogenides, etrc.) and metal-reducers – Mg, Ca, Al, Zn, etc. are used for the synthesis. The advantages of this method are a low price and availability of raw materials. Besides, metallothermal powders are characterized by such valuable properties as high dispersion and homogeneous granulometric composition.

The interaction of the reagents in the combustion wave occurs within two stages. The first one (reduction of the main metal oxide) is a metalthermal reaction. The second stage (SHS itself) is the interaction of the reduced metal with a non-metal followed by a refractory compound formation. There are a lot of secondary reactions which should be suppressed when optimum technological terms of the process are worked out. In the complicated systems of oxide – metal-reducer – carbon (hydrocarbon), carbon-containing components take part in carbide formation and reduction of metal oxides as well. It defines the requirements to the choice of the initial components ratio.

As a result of the SHS with a reducing stage a "semiproduct" is obtained which contains the main compound and the secondary products which can often be metal-reducer oxides. In metallothermic powders the secondary product is distributed uniformly in the whole volume of the reactive mass. So it is necessary to carry out some additional operations to sort out the main compound [24, 25].

Having analyzed the literature data, we can conclude that in the case of the development of the SHS technology of tungsten carbide the main attention should be paid to detection of the terms of nano-particle formation during the synthesis process. However, investigation of the separation methods of chemically pure ultrafine and nanosized compounds from the synthesis products and their analysis are very important too.

2.2. Chemical dispersion

SHS products are cakes or ingots which should be processed for obtaining powders. It can be achieved by either mechanical milling or chemical treatment.

Mechanical milling (conventional milling by balls, friction milling, planetary milling) is the easiest method for obtaining ultrafine and nano-sized powders. It is possible to obtain fine powders (up to 10-20 nm), but the problems of the long duration of the process, powder contamination with the ball and vessel materials, high power consumption require some additional solution.

One of the promising methods of obtaining nano-sized powders is the method microparticle dissolution. Recently, the efficiency of the dissolution processes for converting microparticle size to the nano-level has been confirmed. The method is based on the property of particles to decrease their volume uniformly due to their dissolution in acid and alkali media. But simultaneously the structure and the properties of the central part of the substance or phase remain the same [26].

The main aim of powder application is to obtain a dense product with homogeneous microstructure after compaction. The common reason restricting the refractory material strength is existence of agglomerates in the powder [27]. So in order to make the powder strong, it is necessary to disintegrate or remove large solid agglomerates from the initial powder. In the case of ultrafine powders the agglomerates are disintegrated by dispergating and milling in suitable solutions.

The influence of various solutions on the powder structure, dispersion degree and specific surface area has been already studied for SHS powders of boron nitride and aluminum nitride.

After synthesis, the materials were mechanically disintegrated and subjected to thermochemical treatment in neutral, acid, and alkali media at temperatures ranging from 20 to 100°C [28]. Such treatment is termed "chemical dispersion" of SHS products, as suggested by Merzhanov [29]. Chemical dispersion in a neutral medium resulted in increased total, outer, and inner specific surfaces. Mean grain size decreased. This implies that chemical dispersion provided for disintegration of the materials, as well as leading to formation of new channels and pores and the appearance of new defects, finally resulting in improved specific surface.

In [25] describes thoroughly the application of chemical dispersion for separating ultrafine and nanosized powders of boron nitride obtained by various methods under the SHS mode: from elements, with participation of boron and boron oxide, and from boron oxide with the stage of magnesium reduction.

Possible production of tungsten carbide of ultrafine and nanosized structure by the SHS technology with a reducing stage with using chemical dispersion for separation of submicron powders was of great practical interest.

This paper demonstrates the investigation results of the dependence of SHS tungsten carbide powder dispersion on the SHS process parameters and composition of the solutions used for chemical dispersion of the synthesis products and separation of the final product. The aim is producing single phase tungsten carbide with ultrafine and nanosized structure.

2.3. Experiment description and products characterization

The starting materials used were 99,98+%-pure WO_3 with an average particle size of 10-12 μm (commercially available material which is used in the production of hard alloys), P804-T furnace black less than 45 μm in particle size, and I.PF-1 magnesium powder (99.1+%) ranging from 0.25 to 0.50 mm in particle size.

To mix the components and grind the SHS products, we used ball mills with steel grinding media. Synthesis was carried out in a 30-l SHS reactor under argon atmosphere.

To prepare tungsten carbide, we used the exothermic reaction between tungsten oxide, carbon (black), and magnesium metal:

$$WO_3 + Mg + C + R \rightarrow WC{\cdot}MgO{\cdot}Mg + R' + Q \quad (1)$$

where R is a regulating additive.

The temperature of this process exceeds 3000°C; it can cause decomposition of the forming tungsten carbide. To reduce the combustion temperature, we introduced different additives, inert or decomposing in the combustion wave to form gaseous products. The unstable additives also acted as dispersants ensuring a small particle size of the SHS products.

In addition to tungsten carbide and magnesia, formed in the oxidation-reduction reaction, X-ray diffraction revealed some amount of unreacted magnesium in the intermediate product and also intermediate compounds (magnesium carbides) formed in the synthesis (**Figure 1**).

According to the chemical analysis magnesium content in water-soluble compounds (it should be related to forming carbides) is 0.7 – 0.9 mass %, metal magnesium (unreacted) is 15-17 mass %. The study on the semiproduct microstructure has proved, that ultrafine crystallites of tungsten carbide appear to be embedded into the amorphous phase of the melts of magnesia and metal magnesium (**Figure 2**).

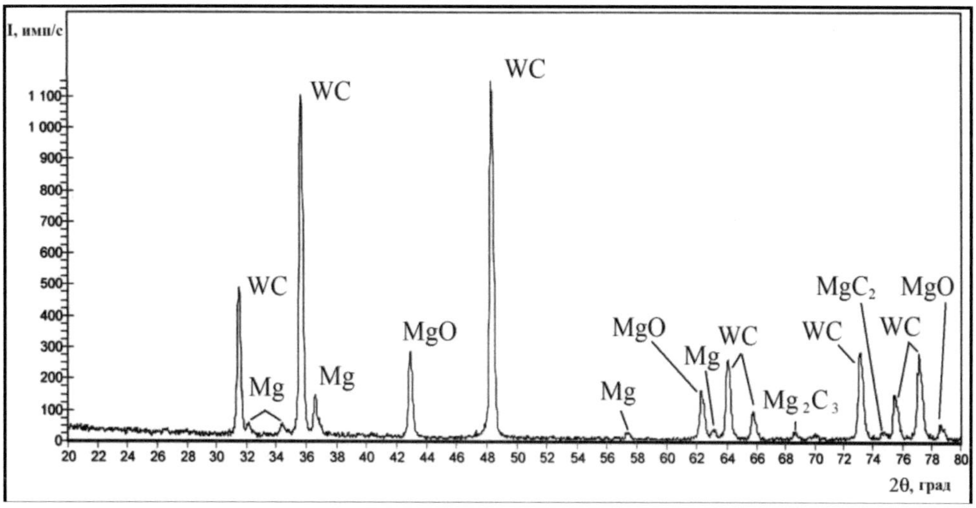

Figure 1. X-ray pattern of WC·MgO·Mg intermediate product.

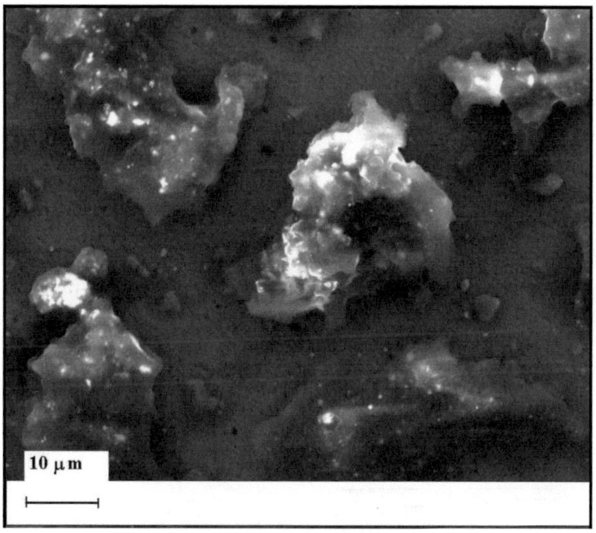

Figure 2. Microstructure of WC·MgO·Mg intermediate product.

The process of chemical dispersion in various solutions is necessary for separation of the target products from the cakes forming during SHS and their further purification from admixtures with simultaneous change in the obtained powder dispersion.

The milled cake was treated with water solutions of hydrochloric acid (1:1) or sulfuric acid (1:5) (acid enrichment) for tungsten carbide separation from the semiproduct. Unreacted metal magnesium and magnesium oxide which was formed during the synthesis process were dissolved.

At first the powder was treated by chloride solutions since it is known that water solutions of haloid salts destroy metal magnesium. Magnesium, potassium and ammonium salts were chosen. It was carried out in order to avoid active gas release when the milled cake was treated with diluted acid solutions (hydrogen release during the interaction of unreacted magnesium with acids) as well as to decrease acid consumption for acid enrichment of the synthesized product.

For decreasing acid consumption, the pulp, consisting of WC·MgO·Mg semiproduct and some amount of magnesium chloride as a catalyst, was saturated with carbon dioxide. During this treatment magnesium content in the solid residue was decreased and in the solution it was increased. Metal magnesium is supposed to transform to solution in the following way:

$$Mg + 2H_2O \rightarrow Mg(OH)_2 + H_2 \quad (2)$$

$$H_2O + CO_2 \rightarrow H_2CO_3 \quad (3)$$

$$Mg(OH)_2 + H_2CO_3 \rightarrow Mg(HCO_3)_2 + 2H_2O \quad (4)$$

At first the pulp is prepared. It is suspension of the treated powder in water. Then the required amount of the acid equal to the stoichiometric ratio is introduced. The addition of water to WC·MgO·Mg is followed by active gas release and the solution heating though distilled water is not supposed to affect metal magnesium greatly due to $Mg(OH)_2$ film appeared on magnesium particle surface [29].

It is known, that at 500°C, MgC_2 can be formed; this carbide is easily disintegrated by water to form acetylene. As the temperature grows from 500 to 600°C, carbon is separated from MgC_2 and Mg_2C_3 appears; this carbide being typical for magnesium only. Methyl acetylene releases during Mg_2C_3 hydrolysis.

So the following reactions can occur in the water solutions:

$$Mg_2C_3 + 4H_2O \rightarrow 2Mg(OH)_2 + HC{=}C{-}CH_3 \quad (5)$$

$$MgC_2 + H_2O \rightarrow Mg(OH)_2 + C_2H_2 \quad (6)$$

$$Mg + H_2O \rightarrow Mg(OH)_2 + H_2 \quad (7)$$

Infrared spectroscopy was used to analyze the gases released in the reaction of WC·MgO·Mg intermediate product with chloride solutions (**Table 1**).

When the intermediate products are treated with potassium chloride and ammonium chloride solutions, a great amount of methane, acetylene, and methyl acetylene is released. It proves the supposition of magnesium carbide formation during SHS. Existence of some amount of methane in the gaseous mixture can be explained by hydrolysis occurring on tungsten carbide particle surface. More gas will be released if ammonium chloride solution is used due to the fact that ammonia is formed during hydrolytic decomposition.

The secondary compounds were removed completely due to the powder treatment with acid solutions

Reactive system	Gas volume, cm^3	Concentration of substance in gas phase, mg/m^3		
		CH_4	C_2H_2	C_3H_4
WC·MgO·Mg +KCl+H_2O	292	16.0	3.1	12.3
WC·MgO·Mg+NH_4Cl+H_2O	405	14.8	1.4	7.2

Table 1. Gas release at WC·MgO·Mg treatment with salt solutions

Figure 3. WC·C powder separated from WC·MgO·Mg semiproduct by acid enrichment

Microstructure analyses (**Figure 3**) have shown, that the tungsten carbide powders resulting from acid enrichment represented large accumulations of fine particles of the main product and unreacted (free) carbon. The chromium mixture (10 g $K_2Cr_2O_7$ in 100 ml H_2SO_4) oxidizes graphite and amorphous carbon at T ≤ 180°C. Preliminary research showed that the treatment of tungsten carbide powder with chromium mixture solution at T ≤ 180°C allowed removing free carbon without dissolving the main product. The carbide powders resulting from acid enrichment were refined with chromium mixture.

As a result, the content of free carbon decreased from 1.0-5.0 to 0.02-0.2%, while the content of oxygen increased due to oxidation of tungsten carbide particle surface. Tungsten carbide particles appeared to be covered by acicular tungsten oxide crystals, which are easily dissolved in diluted alkaline solutions (**Figure 4**).

The changes in the phase and elemental composition of tungsten carbide powder as a result of chemical dispersion in chromic acid mixture and alkaline solutions are presented in **Table 2**.

X-ray diffraction analysis proved that the final products contained only one phase of tungsten carbide. Chemical dispersion in various media caused the primary agglomerates to disintegrate into finer structures of hexagonal tungsten carbide **(Figure 5).**

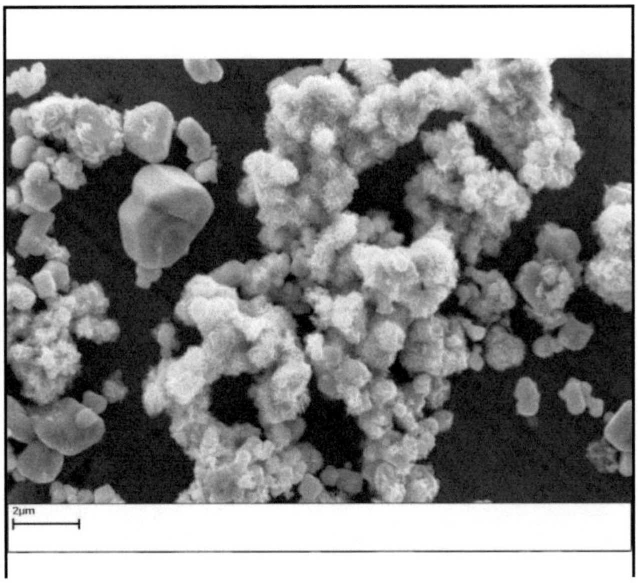

Figure 4. Microstructure of oxidized tungsten carbide powder

Dispersion solution	Dispersion time, h	Weight percent				
		C_{total}	C_{free}	Mg	O	Cr
H_2SO_4 (1:4)	3,0	7,97	2,05	0,063	0,16	
5% $K_2Cr_2O_7$ in H_2SO_4(conc)	3,0	6,12	0,03		0,24	
Aqueous 1% KOH	0,5	6,13	0,03	0,005	0,03	0,005

Elemental analysis of WC·MgO·Mg semiproduct: W_{total} = 44.1 %; C_{total} = 4.1 %; Oxygen = 9.3 % $Mg_{acid.sol.}$ = 37.7 %; Mg_{metal} ~ 15.7 %; $Mg_{water\ sol.}$ = 0.8 %

Table 2. Effect of chemical dispersion on the elemental composition of tungsten carbide powder

3. Results and discussion

The study on SHS stages and chemical dispersion has proved that the final dispersion of the target tungsten carbide product depends on various factors. It was established that the initial mixture composition and density, reactant ratio, their aggregative state in the combustion area, gas pressure, and the nature of regulating additives influenced the size of powder particles.

When calcium chloride or hydride as well as ammonium chloride are used as regulating additives, the final product contains two phases WC and W_2C. When the mixture of ammonium chloride and high-molecular polyethylene or that of metal magnesium and WC·MgO·Mg semiproduct are used, the single-phase target product is obtained.

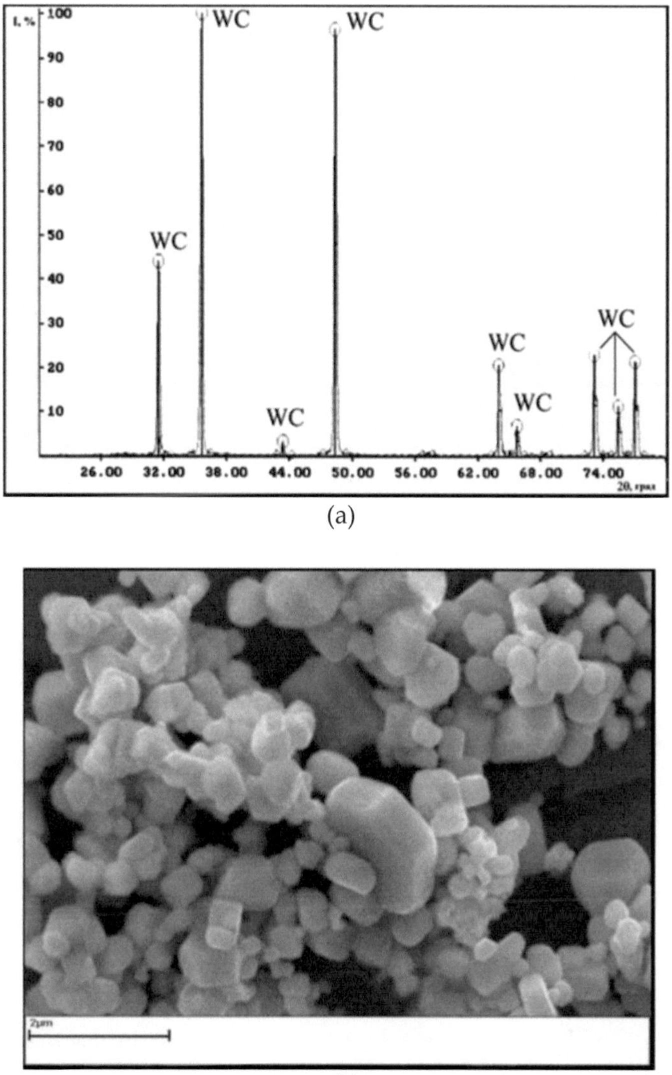

Figure 5. X-ray pattern (a) and microstructure (b) of purified tungsten carbide powder.

The carbon content influenced the phase composition of the product (W_2C content). The single phase product WC is formed in the case of the following ratio of the initial components in the green mixture:

$$33{,}6\%\ WO_3 + 23{,}0\%\ Mg + 2{,}4\%\ C + 40\%\ (WC{\cdot}MgO{\cdot}Mg).$$

The content of magnesium in the starting mixture has a substantial effect on the size of carbide particles: the stoichiometric amount of magnesium results in coarse powders, while it excess leads to a fine product (**Figure 6**).

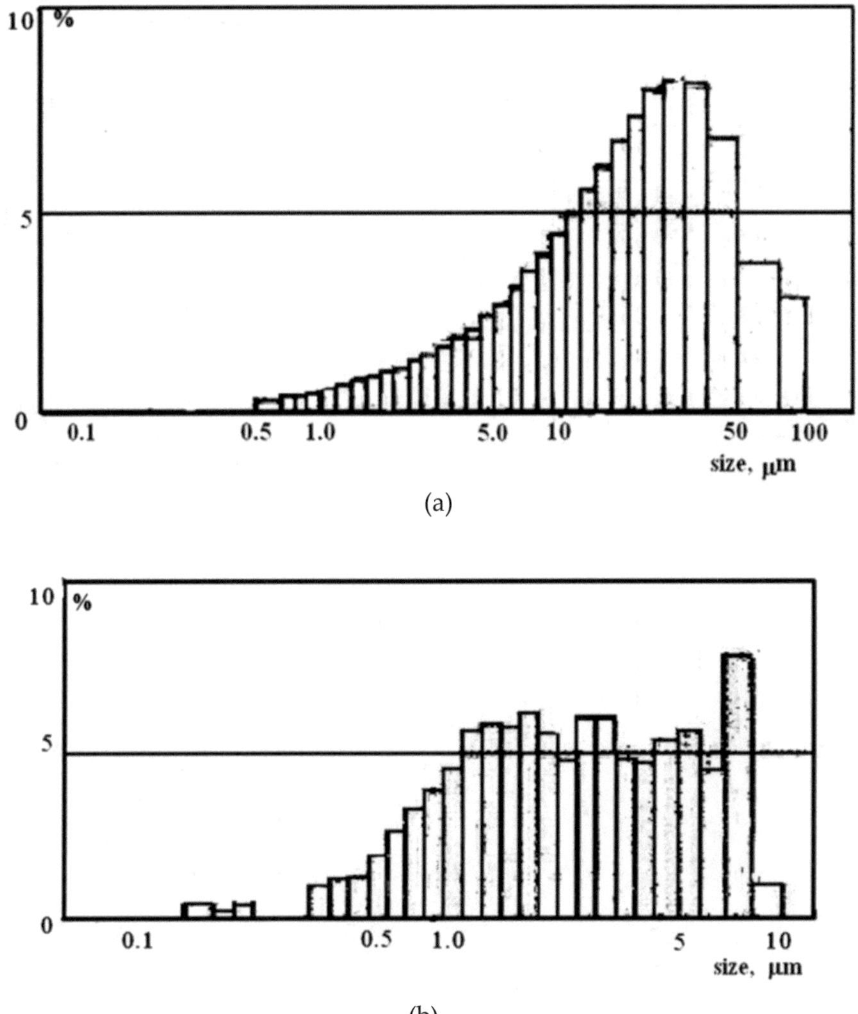

Figure 6. Particle size distributions in tungsten carbide powders: **(a)** stoichiometric amount of magnesium in the starting mixture, **(b)** excess of magnesium in the starting mixture.

The excess of magnesium in the mixture seems to inhibit the growth of tungsten carbide crystals and to form a liquid phase when carbides are crystallized; the liquid phase and adjusting additives prevent intensive crystal growth. Introduction of WC·MgO·Mg into the green mixture also decreases the dispersion degree of the final product. Probably, the introduced additives as well as metal magnesium form a liquid phase under the terms of crystallization. Tungsten carbide ultrafine crystals contained in the introduced semiproduct can accelerate tungsten carbide crystallization and appear to be crystallization centers but a rather viscous medium prevents intensive crystal growth. Coating of tungsten carbide particles with liquid melt results in better stability of tungsten carbide to hydrolysis and oxidation after the synthesis process.

In studying chemical dispersion, the above results were used to analyze how the composition of the solutions, used to recover tungsten carbide from synthesized products, influenced the structure and particle size of the final tungsten carbide powders. The following systems were used:

- diluted sulfuric acid (1 : 5),
- diluted hydrochloric acid (1 : 1),
- ammonium chloride and hydrochloric acid solutions,
- potassium chloride and hydrochloric acid solutions.

It was established, that the tungsten carbide particle size depends on the composition of solutions used at the first chemical dispersion stage: recovery of carbide from intermediate product (**Table 3**).

Acid enrichment conditions	Volume fraction, %	
	≤ 300 nm, %	≤ 500 nm, %
HCl (1:1)	61,3	87,5
H_2SO_4 (1:5)	66,7	86,3
30 %NH_4Cl + HCl	81,6	96,5

Table 3. Fraction volumes of refined tungsten carbide powders with minimum particle sizes.

This result can be explained by the following way. Tungsten carbide is thermodynamically unstable and can be oxidized in the medium of water or oxygen at the room temperature [30]. X-ray phase analyses of tungsten carbide powder state in the humid medium show, that the surface of tungsten carbide particles is the first to be oxidized. The thickness of the oxide film increases with an increase in humidity [31].

In water the oxide film is entirely removed due to its dissolution and formation of tungstate ions by the reaction:

$$WO_3+H_2O \rightarrow WO_4^{2-}+2H^+ \quad (8)$$

When the milled semiproduct is dispersed by ammonium chloride or potassium chloride solutions, the pH of solution changes from low acid values to high alkali ones. The forming medium provides acceleration of oxide film dissolution by Reaction 8 and deeper tungsten carbide particle hydrolysis leading to a decrease in the particle size due to dissolution from the surface. So, chloride application at the stage of acid enrichment allows obtaining tungsten carbide powder with the number of particles of less than 300 nm in size being 80 % of the total number **(Figure 7)**. Using suitable emulsifiers can disintegrate the agglomerates and separate tungsten carbide particles of less than 100 nm from ultrafine ones.

Application of ultrasound in the process of chemical dispersion decreases the time of the process and affects the dispersion degree of the product. In the case of mechanical mixing refining of tungsten carbide powders with chromium mixture takes several hours. The ultrasound effect decreases the time to 30 – 40 minutes. It can be explained by disintegration of tungsten carbide agglomerates and carbon coarse particles and acceleration of the reduction-oxidation reaction of chromium mixture with free carbon.

The ultrasound effect on tungsten carbide composition and dispersion has been studied (**Table 4**).

Refinement time	Refinement temperature	C_{total}, mass %	C_{free}, mass %	Oxygen, mass % (non-purified product)	Oxygen, mass % (purified product)
30 min	145°C	5,72	0,015	1,40	0,14
45 min	85°C	5,18	0,013	2,35	0,07

Table 4. Ultrasound effect on tungsten carbide powder composition at final product refinement

After refining with chromium mixture, the carbon content decreases to ~0.1 % but oxygen content increases greatly (in comparison with mechanical mixing) due to oxidation of tungsten carbide particle surface. The lower the refinement temperature and the higher time of ultrasound action are used, the higher dispersion of tungsten carbide powder is achieved (**Figure 8**). Under these terms the process of tungsten carbide particle surface oxidation is more active; therefore the particle size is actively decreased (powder A). An increase in the refinement temperature results in obtaining less dispersed powder B due to dissolution of fine particles under the strict terms of the process.

The powder **(a)** consists of agglomerates of fine and coarse particles. It is possible to separate ultrafine and nanosized tungsten carbide particles using proper technological terms. In the powder **(b)** fine tungsten carbide particles are situated on the surface of coarser particles and it makes their further separation much more difficult. Therefore, the ultrasound application results in additional milling of tungsten carbide powders and more complete purification from admixtures.

The results of the work on SHS of tungsten carbide powder with the reduction stage led to the development of the industrial technology of ultrafine and nanosized tungsten carbide powders synthesis. **Figure 9** demonstrates the curve of the particle size distribution of tungsten carbine powder synthesized in the industrial reactor. Obviously, the product is a mixture of particles of different sizes. The prevailing particles are ultrafine and nanosized ones.

Tungsten carbide powders synthesized by the developed technology were tested in making alloys and items thereof.

We studied sinterability of fine-particle of SHS tungsten carbide powders. **Table 5** compares the physicochemical properties and structure of WC-Co alloy prepared with the use of SHS tungsten carbide and the commercial alloy VK6-OM (containing tungsten carbide produced by a furnace process).

(a)

(b)

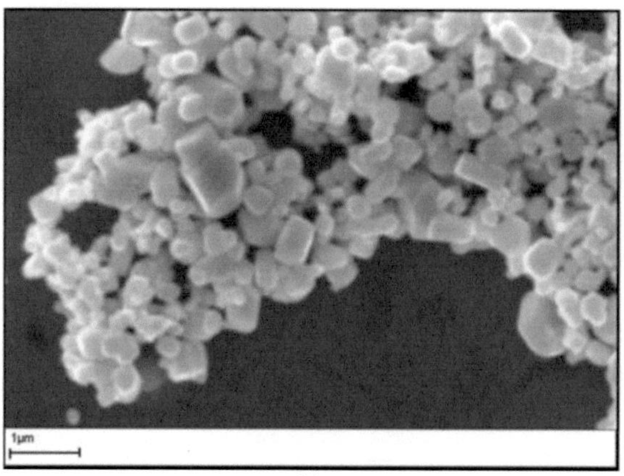

(c)

a - HCl (1:1); **b** – NH_4Cl (30 % solution) + HCl (1:1); **c** - KCl (30 % solution) + HCl (1:1)

Figure 7. Tungsten carbide powder microstructure depending on the terms of acid enrichment

Figure 8. Dependence of refined tungsten carbide powder microstructure on the terms of ultrasound treatment: A – T=85ºC; B – T=145ºC.

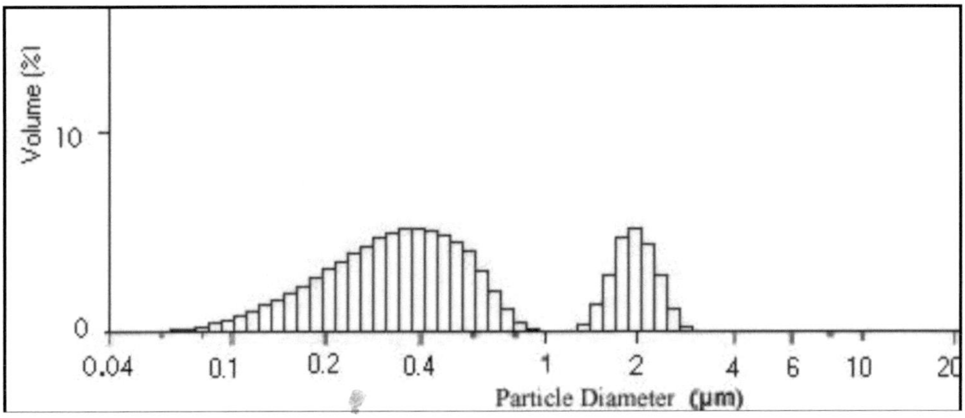

Figure 9. Particle size distribution of tungsten carbide powder synthesized in industrial reactor.

Parameters	SHS WC alloy	VK6-OM alloy (standard)
Density, g/cm^3	14.9	14.7
Hardness, HRA	91	90.5
Coercivity, A/m	270	280
Bending strength, σ, kg·f/mm^2	170	130
Durability coefficient	1.4	1.0
Porosity, A %	0.04	0.2
Percentage of particles less than 1 µm in size	80 %	60 %

Table 5. Physicochemical properties of WC-Co alloys prepared by using WC-SHS and WC-furnace process.

The bending strength, durability coefficient, and dispersion degree of the alloy produced from SHS tungsten carbide exceed those of the commercial alloy.

As a result of the realized research, the technology of Self-propagating High-temperature Synthesis has been developed and is being introduced for production of ultrafine and nanosized tungsten carbide powder with the use of chemical dispersion for separation, purification and additional milling of the target product.

Organization of industrial SHS production of submicron tungsten carbide powders includes:

- development of hydrometallurgical stage of submicron tungsten carbide powder separation;
- development of the production line with complete or partial automation;
- organization of design work in modernization of non-standard equipment and in selection of standard additional devices;
- preparation of the workshop for tungsten carbide semiproduct treatment (leaching, utilization and regeneration of wastes).

The annual production output is 150 tons. The profitableness is up to 80 %.

4. Conclusion

The processes of Self-propagating High-temperature Synthesis were studied for obtaining nanosized powders of refractory compounds, particularly, tungsten carbide. The SHS terms influence crystallization of the obtained powders. Varying the SHS parameters (reactant ratio, regulating additives, inert gas pressure, combustion and cooling velocities) allows changing tungsten carbide powder morphology and particle size.

SHS tungsten carbide powder differs from its furnace and plasmochemical analogs in structure and purity. The grain size can be governed during the SHS processes. Powders of less than 100 nm in particle size can be obtained at complete suppression of recrystallization in combustion products. Separation of the powders from the milled cakes by chemical dispersion with various solutions and choice of chemical dispersion terms (the solution composition, the process time and temperature) allow obtaining SHS materials with the nanostructure characterized by high specific surface area and particle size less than 100 nm with simultaneous preserving the phase and chemical composition of the product.

As a result of the realized research, the technology of Self-propagating High-temperature Synthesis has been developed for production of ultrafine and nanosized tungsten carbide powder with the use of chemical dispersion for separation, purification and additional milling of the target product. The sinterability of the synthesized tungsten carbide powder was studied. The bending strength, durability coefficient, and dispersion degree of WC-Co alloy produced from SHS tungsten carbide exceed those of the commercial alloy.

The proposed technology of ultrafine and nanosized tungsten carbide powder synthesis has some advantages in comparison with the available technologies:

- Availability of theoretically explained backgrounds for governing the reaction temperature and velocity and component conversion completeness, which provide the possibility of obtaining high quality products of the preset structure at optimum terms;
- Low requirements to the initial mixture quality since partial self-purification of SHS products from admixtures takes place during the combustion process;
- Simple equipment using various approaches of physical influence on the substance;
- Possibility of industrial production of nanosized materials.

Nowadays, the number of ultra-dispersed materials produced in industry is restricted. Development of industrial production technologies and widening of application fields of nanosized materials is commercially important.

Author details

I. Borovinskaya, T. Ignatieva and V. Vershinnikov
Institute of Structural Macrokinetics and Materials Science, Chernogolovka Moscow Russian

Acknowledgement

The authors wish to express their deep thanks to Academician A.G.Merzhanov for useful comments and suggestions.

5. References

[1] Rafaniello W (1997) Critical powder characteristics. In: Weimer A, editor. Carbide, nitride, and boride materials synthesis and processing. London: Chapman æ Hall. 671 p.

[2] Schubert W, Bock A, Lux B (1995) General aspects and limits of conventional ultrafine WC powder manufacture and hard metal production. Int. J. Refract. Metals Hard Mater. 13: 281-296.

[3] Jia K, Fischer T, Gallois B (1998) Microstructure, hardness and toughness of nanostructured and conventional WC-Co composite. Nanostruct. Mater. 10: 875-891.

[4] Spriggs G (1995) A history of fine grained hard metal. Int. J. Refract. Met. Hard Mater. 13: 241-255.

[5] Cottrell A (1995) Chemical bonding in transition metal carbides. London: The Institute of Materials. 97 p.

[6] Tägtström P, Högberg H, Jannson U, Carlsson J.-O (1995) Low Pressure CVD of Tungsten Carbides. J. de Phys. IV. 5: 967-974

[7] Schwartzkopf P, Kieffer R (1953) Refractory Hard Metals: Borides, Carbides, Nitrides and Silicides. New York: The MacMillan Company. 447 p.

[8] Rieck G (1967) Tungsten and its Compounds. Oxford: Pergamon Press. 138 p.

[9] Brookes K (1992) World Directory and Handbook of Hard Metals and Hard Materials. Hertfordshire UK: Int.Carbide data. p.88

[10] Gao L, Kear B (1995) Low Temperature Carburization of High Surface Area Tungsten Powders. Nanostruct. Mater. 5: 555-569.

[11] Gao L, Kear B (1997) Synthesis of Nanophase WC Powder by a Displacement Reaction Process. NanoStructured Mater. 9: 205-208.

[12] Medeiros F, De Oliveira S, De Souza C, Da Silva A, Gomes U, De Souza J (2001) Synthesis of WC through gas-solid reaction at low temperature. Mater.Sci.Eng. A 315: 58-62.

[13] Bock A, Zeiler B (2002) Production and characterization of ultrafine WC powders. Int. J. Refract. Metals Hard Mater. 20: 23-30.

[14] Krishna B, Misra V, Mukherjee P, Sharma P (2002) Microstructure and properties of flame sprayed tungsten carbide coatings. Int. J. Refract. Metals Hard Mater. 20: 355-374.

[15] Merzhanov A, Shkiro V, Borovinskaya I (1971) Synthesis of refractory inorganic compounds. Certif. SSSR No. 255221 Appl. N 1170735. Byull. Izobr. N 10.

[16] Merzhanov A, Borovinskaya I (1972) Self-propagating high-temperature synthesis of refractory inorganic compounds. Dokl. AN SSSR 204: 366-369. (in Russian)

[17] Merzhanov A (1991) Advanced SHS ceramics: today and tomorrow morning. In: Soga N, Kato A, editors. Ceramics: Toward the 21st Century. Tokyo: Ceram. Soc. Jap. Publ. pp.378-403.

[18] Merzhanov A (1992) New manifestation of an ancient process. In: Rao C, editor Chemistry of Advanced Materials. Blackwell Sci. Publ: pp. 19-39.

[19] Mukasyan A, Borovinskaya I (1992) Structure formation in SHS nitrides. Int.J.of SHS. 1: 55-63.

[20] Shugaev V, Rogachev A, Merzhanov A (1993) Structure formation of SHS products in model experiments. Inzh. Fiz. Zh. 64: 463–468. (in Russian)

[21] Merzhanov A (1984) Macroskopic kinetics and modern chemistry. Proc. I All-Union conference on macrokinetics and gas dynamics. Alma-Ata

[22] Borovinskaya I (2003) SHS Nanomaterials. In: Merzhanov A, editor. SHS: Concepts of Current Research and Development. Chernogolovka: Territoriya. pp. 178-182. (in Russian)

[23] Merzhanov A (1990) Self-propagating high-temperature synthesis: twenty years of search and findings. In: Munir Z, Holt J, editors. Combustion and plasma synthesis of High-temperature Materials. New York Press: VCH PubL. pp. 1-53

[24] Lagunov Y, Pikalov S, Kolomeets G, Mamyan S(1981) Boron nitride fabrication by SHS-product enrichment with reduction stage. In: Merzhanov A, editor. Technological Combustion Problems. Chernogolovka, pp.40-42. (in Russian)

[25] Borovinskaya I, Ignatieva T, Vershinnikov V, Khurtina G, Sachkova N (2003) Preparation of Ultrafine Boron Nitride Powders by Self-propagating High-Temperature Synthesis. Inorg. Mater. 39: 698-704.

[26] Lee C.-S, Lee J.-S, Oh S.-T (2003) Dispersion control of Fe_2O_3 nanoparticles using a mixed type of mechanical and ultrasonic milling. Mater. Letters. 57: 2643– 2646.

[27] Lange F (1989) Powder processing science and technology for increased reliability. J.Am.Ceram.Soc. 72: 3-15.

[28] Borovinskaya I, Vishnyakova G, Savenkova L (1992) Morphological features of SHS boron and aluminum nitride powders. Int. J. of SHS. 1: 560-565.
[29] Remy H (1960) Lehrbuch der anorganischen Chemie. Leipzig: Akademische verlagsgessellschaft geest & Portig K.-G. B. 1, 900 S.
[30] Warren A, Nylund A, Olefjord I (1996) Oxidation of tungsten and tungsten carbide in dry and humid atmospheres. Int. J. Refr. Metals Hard Mater. 345–353.
[31] Webb W, Norton J, Wagner C (1956) Oxidation studies in metal–carbon systems. J. Electrochem. Soc. 112–117

Chapter 2

Spark Plasma Sintering of Ultrafine WC Powders: A Combined Kinetic and Microstructural Study

A.K. Nanda Kumar and Kazuya Kurokawa

Additional information is available at the end of the chapter

http://dx.doi.org/10.5772/55291

1. Introduction

Nano grained cemented tungsten carbide (n-WC) is currently being researched for many potential applications in manufacturing processes. An example is the near net shape manufacturing of aspheric glass lenses. With the advent of optical technology and electro-optic systems, conventional spherical lenses are now being replaced by aspheric lenses of smaller dimensions and lower curvatures to be accommodated inside flat cellular phones and DVD readers. A cost effective method of fabricating such small aspheric lenses is by molding the glass gob in a suitable preform or mold at temperatures near the glass transition temperature (T_g). WC-based cemented carbides are a natural choice for the mold because of their high hot hardness and low coefficient of thermal expansion, CTE (which is compliant with the CTE of glass). A major issue in this near net shape fabrication method is that the surface finish of the carbide mold should be extremely smooth as otherwise the glass component will also reproduce the surface roughness of the mold. This eventually leads to aberration of the lens and a loss of precision, consequently necessitating the need for an extra grinding or polishing step after the manufacturing process. Ultra-fine grained carbides, owing to their small grain size, can be polished to extreme smoothness of the order of 2-3 nm or lesser. To facilitate the lens' release from the mold, usually Ir or Re coatings are applied on the mold surface. Generally, this arrangement works well for near net shape mass production of small aspheric lenses and is commonly used in lens manufacturing industries. Another instance where n-WC assumes commercial importance is in the micromachining industry where often extremely small holes have to be drilled into hard substrates. The drill-bit in such applications is made of WC with a very small curvature at its tip which is possible only if the grain size is in the nano-metric range. Larger grains lead to blunting when the tip undergoes brittle intergranular fracture resulting in chipping off a large chunk of the material from the drill tip.

Given that cemented *n*-WC has many such industrial applications particularly owing to its mechanical strength, the microstructure, porosity (density) and grain size inarguably are of extreme significance in tailoring its properties like hardness, toughness and chemical stability. Powder metallurgical processes like Hot Iso-static Pressing (HIP) and high temperature solid state or liquid phase sintering are the usually employed methods of fabricating dense compacts of pure or cemented WC. However, pure WC in the absence of a binder is rather difficult to consolidate completely. While in cemented WC, the liquid phase assists sintering by particle rearrangement, the low diffusivities of W and C under pure solid state sintering conditions retard quick consolidation during sintering or HIP of pure n-WC. Therefore, unnaturally long durations (in the case of isothermal sintering) or very high temperatures in excess of 2000 C (in the case of non- isothermal sintering) are required for consolidation of n-WC. This disadvantage has led researchers to seek alternate or improvised sintering methods [Bartha L *et al*, 2000, Agrawal D *et al*, 2000, Breval E *et al*, 2005, Kim H C *et al*, 2004] like Spark Plasma Sintering (SPS) or microwave sintering to achieve quicker densification at lower time costs. The SPS method, in particular has attracted wide attention owing to its consistently good record of achieving the desired density at surprisingly low times and lower temperatures. The generation of very high current densities leading to a sort of, 'plasma welding' between the particles is suspected to be the chief cause of such a profit in the total energy budget compared to conventional sintering. However, no clear evidence exists for the actual generation of plasma or any surface melting phenomenon in the SPS process although the hypothesis has been widely debated [Tokita M, 1997, Hulbert D M *et al*, 2008, Hulbert D M *et al*, 2009].

Since the last decade, a number of reports on SPS of *n*-WC have consistently come up in journals and scientific magazines. Not only have the compacts been manufactured to complete density, but the grain size could also be limited to the ultra-fine size (200-400 nm). Usually Hall-Petch hardening is observed at low grain sizes and low cobalt content. This increased capability to constrain the microstructure to the ultrafine regime has been largely aided in part because of the commercial availability of nano powders of WC synthesized by many chemical routes and also partly because of the current popularity of activated sintering instruments that also accommodate high heating rates and pressure along with the presence of electromagnetic fields.

2. Activated sintering processes

Sintering methods involving the presence of an electric field are generally called Field Assisted Sintering Techniques (FASTs). Unlike conventional sintering - in which the sample is heated from the outside (furnace) - in FAST, the sample is heated internally by the passage of an electric current. Compared to the hot pressing process, FAST methods can have extremely high heating rates, sometimes even upto 2000 K/min [Tokita M *et al*, 2007, Cramer G D, 1944 and a host of other patents, a review of which can be found in the paper by Salvatore Grasso *et al*, 2009]. This is achieved by using current pulses from a few micro seconds to milli seconds but charged with an extremely high current density of about 10,000 A/cm^3. External pressures can also be applied from a few MPa to typically 1000 MPa making the sintering process rapid and effective. Generally, the electric field can be applied in a number of ways: pure DC (also called resistive sintering), pulsed DC or Microwave. Activated sintering using a pulsed DC has also been often referred to as Spark Plasma Sintering (SPS) in the literature, since the high current density is thought to induce a plasma at the inter-particle neck region. However, the generic term, Pulsed Electric Current Sintering (PECS) is also commonly used in reference to any type of current waveform other than pure DC.

In a typical SPS process, the powder sample is loaded in a cylindrical die and closed on the two sides by electrically conductive punches. For ease of separation after sintering and also to avoid any reaction between the punch and the sample, graphite papers are used as spacers. Sintering is carried out in vacuum and both pressure and electric current through an external power source is applied to the punches. The electric field control can be achieved in two ways: in the *temperature controlled* mode, the current to the punch and sample is supplied according to a pre-set temperature programme. The temperature is measured at the die surface with a pyrometer and the feedback is used to adjust the current supply accordingly. In the *current controlled* mode, a constant current is supplied to the sample and the temperature is monitored. Very high heating rates can be achieved limited only by the maximum current available from the power source. However, the actual temperature in SPS can be quite different from the measured temperatures for many reasons: the pyrometer measures the temperature at a niche in the die which is neither exactly on the sample surface nor in the surface interior - certain reports put this difference at ~50-100 K [Bernard and Guizard, 2007]; measured temperatures are usually the average values and give no indication of the very local temperatures that can actually exist between the particles. The overall electrical resistance - including the internal resistance of the voltage source and resistance of the bulk of the apparatus - controls the current flow and consequently, the Joule heating generated in the sample. Hence, in an SPS experiment, the total resistance, R_{total} can be written as:

$$(R)_{total} = (R)_{internal} + (R)_{contact} + (R)_{sample} + (R)_{bulk} \tag{1}$$

It has been found that for a constant applied current, the maximum resistance (and thereby the maximum joule heating) occurs at the punch/graphite contact surface, $R_{contact}$ [Giovanni Maizza *et al*, 2007, Munir Z A et al, 2006]. Moreover, the resistance of the sample, R_{sample} is continuously changing (as a function of the instantaneous porosity) and hence, the observed value of current in circuit is a product of the complex interplay of various parameters. The pulse frequency of the DC supply in a typical SPS process is split into an ON/OFF ratio of 12/2. The ON pulse in turn is split into sub pulses of milli second duration. All these parameters can be controlled by the user to achieve the best sintering conditions. Usually, only the heating rate and pressure are varied with the rest of the controls kept according to the factory settings.

While the quantum of publications on/using SPS has been steadily increasing, the basic process is far from being well understood; the answer to the fundamental question of whether a plasma is generated at the inter-particle contact area is still elusive. Another intriguing fact is the observation of very low sintering activation energies, enhanced sintering rates and low sintering temperatures when the sample is subjected to a simultaneous pressure and electric field as in SPS. While some authors attribute this observation to electro-migration (i.e., diffusion under an electric field gradient) as a, 'sintering enhancer', it must be noted that electro-migration can be expected to play a serious role in the sintering of highly ionic compounds. But the observation that the activation energy can be equally low in predominantly covalent compounds like WC (the ionicity according to the Pauling scale is only ~1%) suggests that the field effect may not be the sole cause for the observed rapid kinetics. Thermodynamic arguments suggest that the applied pressure drives sintering while the electric field retards grain growth thereby achieving full densification with limited grain growth. A number of alternate mechanisms, which treat the GB as a separate phase have also been put forth [Dillon S J *et al*, 2009, Di Yang *et al*, 2010, Gupta V K *et al*, 2007]. However, while the outcome has been certainly encouraging, a clear and validated picture of the sintering mechanism under activated sintering is still lacking.

3. Isothermal and non-isothermal sintering

Sintering, like coarsening and grain growth is also a thermally activated process and hence an Arrhenius type of dependence on temperature is observed. The kinetics of fusion of two particles during sintering is usually studied either by measuring the neck to particle size ratio (x/a) or by measuring the macroscopic shrinkage using a dilatometer with respect to time. A number of theories have been developed to explain both shrinkage and neck growth during sintering [Ashby M F, 1974, Swinkels F B and Ashby M F, 1981, Beere W, 1974, Coble R L, 1958]. Such theories derive explicit relations connecting the shrinkage strain, $\varepsilon\,(=\Delta l/l_0)$ or neck growth (x/a) to the time of sintering, t under isothermal conditions. Measurements of neck growth in ultrafine particles are difficult and therefore, the macroscopic shrinkage strain is instead measured and a suitable theory is chosen to study the kinetics. In any case, the sintering kinetics (either solid or liquid phase assisted) can be described by a generic equation of the type:

$$\varepsilon^m = \left(\frac{\Delta l}{l_0}\right)^m = \frac{Kt}{T} \qquad (2)$$

where 'm' is the sintering exponent, t is the isothermal holding time and T is the hold temperature. The higher the value of m, the lower is the magnitude of shrinkage. The constant, $K = K(T)$ is the temperature dependant sintering constant and accommodates the interface energetics and transport kinetics of the sintering process via the surface energy, γ and the diffusion coefficient, D. The form of K can be related to temperature by an Arrhenius type equation,

$$K = K_0 e^{-Q/RT} \qquad (3)$$

where Q refers to the activation energy for densification and R is the gas constant. The kinetic parameters can be evaluated easily by a simple modification of the two equations. Firstly, equation (2) gives:

$$ln(\varepsilon) = \frac{1}{m} ln\left(\frac{K}{T}\right) + \frac{1}{m} ln(t) \tag{4}$$

Therefore a plot of ln (ε) against ln (t) at constant T is a straight line with slope $1/m$. The sintering exponent 'm' can vary depending on the mechanism (diffusion path) and geometry of the sintering bodies. **Table 1** shows the various values of m available in the literature, modelled for the sintering of a pair of spherical particles.

The activation energy for sintering, Q can be determined in many ways: Utilizing the exponential dependence of K on T, and the m value determined earlier, we can write,

$$ln(T\varepsilon^m) = ln(K_0 t) - \frac{Q}{RT} \tag{5}$$

Hence a plot of ln ($T\varepsilon^m$) against $1/T$ at constant values of time, t should yield a straight line from which Q can be determined if the sintering exponent, m is known. Another equivalent method for determination of the activation energy of sintering in isothermal experiments is the time for constant fraction technique which is based on the measurement of a constant linear shrinkage fraction at different hold temperatures. The activation energy can then be determined by:

$$ln(t) = ln\left(\int_0^f [k_0 f(\varepsilon)]^{-1} dy\right) + \frac{Q}{RT} \tag{6}$$

where $f(\varepsilon)$ is the shrinkage strain – time curve.

A more common method of determining the activation energy without *apriori* knowledge of the sintering exponent, m is the Dorn's method [Bacmann J J and Cizeron G, 1968]. Here, the densification strain rates are evaluated at a constant time at different sintering temperatures so that the slope of a plot of ln(dε/dt) against $1/T$ would yield values of Q. Usually the Dorn method is associated with an error of ~8 to 10%. Provided the initial temperature instability during the first few minutes of isothermal hold is eliminated and if the system does not exhibit shrinkage saturation (asymptotic behaviour) very early during the hold period, both kinetic methods should yield the same values of activation energy.

Equations (2)-(6), hold only during the initial stages of sintering. At later stages of sintering, the free energy reduction accompanying grain growth exceeds that of neck growth. When neck formation is succeeded by interconnected pore structures, the intermediate stage is said to have started. This stage is usually reached after the compact attains 80% or greater of the final density. Compared to the initial stage, fewer models are available for this stage owing to two primary reasons: complicated pore/particle geometry and concurrent grain growth. Densification strain equations for the intermediate stage are primarily based on pore/particle geometries and the inter-relation between them. The frequently referred intermediate stage model is the tetrakaidecahedron model of Coble [Coble R L, 1961a, Coble R L, 1961b]. The appropriate shrinkage kinetics is derived in terms of porosity (pore fraction) rather than linear shrinkage and expressed for different mechanisms as follows:

Lattice diffusion without grain growth

$$P - P_0 = \frac{N_A D_v \Omega \gamma}{k_B T G^3} t \tag{7}$$

Lattice diffusion with grain growth

$$P - P_0 = \frac{N_A D_v \Omega \gamma}{k_B T G^3} ln(t) \tag{8}$$

Grain boundary diffusion without grain growth

$$P - P_0 = \left(\frac{N_A D_b w \Omega \gamma}{k_B T G^4}\right)^{2/3} t^{2/3} \tag{9}$$

where the terms have the following meanings: P_0 – initial porosity at $t = 0$ in the intermediate stage, P – final porosity, D_v, D_b – volume, grain boundary diffusivities, γ– surface energy, w –grain boundary width, Ω– atomic volume, G – grain size and the other terms have the usual meanings.

Non isothermal (also called *constant rate of heating*, CRH) sintering can also be analysed by suitable models. In this work, we employed the method of Young and Cutler [Young W S and Cutler I B, 1970] to determine the activation energy from a plot of ln(dε/dt) against 1/T. The slope determined from the plot is mQ (effective activation energy) and if either the mechanism (m is ½ for LD and ⅓ for GB diffusion) or activation energy (Q) is known *apriori* (from isothermal experiments), the other unknown can be determined. We used a combination of both isothermal and non-isothermal sintering to complement each for the kinetic studies reported in this work.

Diffusion pathway	Value of *m*	Reference
LD	0.46	Johnson and Cutler, 1963a
	0.5	Coble R L, 1958
	0.4	Kingery W D and Berg M, 1955
GB	0.31	Johnson and Cutler, 1963b
	0.33	Coble, 1958

Table 1. Values of the initial stage sintering exponent developed for model geometries. (LD and GB refer to lattice diffusion (i.e., volume) and grain boundary respectively).

4. Experiments

Commercially purchased *n*-WC powders without any pre-treatment were used for sintering. The particle size measured by BET was 70 nm and the powder composition included 0.4% O, 5 ppm Cr, 27 ppm Fe, 4 ppm Mo, 3 ppm Ca, 2 ppm Ni, <5 ppm Si and < 2 ppm Sn. Approximately 2.5 – 3 g of the powder was filled into a 10 mm diameter graphite die for spark plasma sintering (SPS) in a Dr SINTER LAB instrument. This SPS instrument has a dilatometer with an accuracy of 0.01 mm for measuring the instantaneous linear shrinkage. Temperature measurements were carried out using a radiation thermometer (pyrometer) that was focused on a small niche in the carbon die. Graphite sheets were used as spacers to separate the powder sample from the punch and die. After initial temperature stabilization at 873 K for 3 minutes, sintering was carried out in vacuum (< 4 Pa) at a constant heating rate of 50 K/min and a compressive stress of 40 MPa to various temperatures from 1073K to

1873 K. The samples were held at these temperatures for a period of 30 minutes while their shrinkage was continuously monitored using a dilatometer. For the non-isothermal sintering studies, two heating rates – 20 K/min and 50 K/min – were employed and the sintering process was assumed to be complete when the dilatometer showed no further change in shrinkage during two successive temperature measurements. All the samples were allowed to cool down to room temperature inside the chamber. Before analysis, the samples were first polished with fine diamond paste (1μm) and subsequently cleaned with ethanol in an ultrasonic bath. The densities of the samples were determined by the Archimedes method. All densities are reported relative to the density of pure WC (15.8 g/cc). Fractured and etched samples were used for the microstructure analysis. Before etching, the samples were cross sectioned, polished and cleaned as earlier. Conventional Murakami solution (H_2O+KOH+$K_3[Fe(CN)_6]$ in a volumetric ratio of 10:1:1) was used for etching the compacts. For TEM analysis, the cross sectioned samples were mechanically thinned to 100 μm, dimpled to a depth of 20 μm and then milled with Ar ions to electron transparency. Microstructure and phase analyses were carried out using XRD, FE SEM, EBSD and TEM. Grain size evaluation was performed using the FE SEM images (15000 X magnification) of the etched samples with the aid of an image analysis software (Image Pro-Plus). Approximately 150-200 grains from three different locations of a sample were randomly selected for the measurements. The boundaries were delineated either manually or auto segmented and the average diameter (average value of the diameters measured at 2° intervals and passing through the centroid of the selected grain) of the grains was calculated.

5. Results

5.1. Analysis of the sintering kinetics

Fig.1 shows the combined isothermal and non-isothermal shrinkage curves. The immediate point worthy of interest is that the CRH strain rate curve does not exhibit a unimodal, gaussian type behaviour that is generally observed in the non-isothermal sintering of many ceramics [Wang J and Raj R, 1990, Panda *et al*, 1989, Raj R and Bordia R K, 1984]. Instead, there are two peaks (at around 1450 K and 1900 K) leading to a broad plateau covering a rather large temperature interval (from approximately 1400 K to 1900 K). At the peak points in the CRH curve, the corresponding isothermal curves also show a large increase in strain which varies proportionally with the relative magnitude of the CRH sintering strain rate; in most of the low temperature regime, the isothermal sintering strains show saturation, implying that the sintering strains are critically dependant on the heating rate and the temperature of isothermal hold. In conventional sintering, the heating rate is usually assumed to be irrelevant to the kinetics as the sample is presumed to reach the isothermal sintering temperature very swiftly. Our comparison shows the explicit dependence of the isothermal curves on the non-isothermal sintering trajectory and sintering temperature. These preliminary results confirm that the sintering behaviour is not governed by a simple, single mechanism. In the same **Fig.1**, the stages are marked as Initial, Intermediate I and II for ease of analysis. Although the curve does not resemble the typical three stage sintering process, it does indeed show at first glance, the occurrence of sub-stages.

As mentioned in the previous sections, the relevant equations of sintering have to be applied only to the corresponding sintering stages. Delineating a particular sintering stage (initial, intermediate or final) can be carried out by real time observation of the microstructure. However, such a process is tedious and quite ambiguous, particularly if the particle size is of the order of a few tens or hundreds of nm. As a general rule, when the measured linear shrinkage strains are less than 5%, the dynamics can be assumed to be in the initial stage. With this presumption, the subsequent analysis was carried out for the temperature range 1073-1273 K. Linear shrinkage strains and calculated sintering exponent in the initial stage are shown in **Fig. 2a,b**. Clearly, while the net shrinkage strains are less than 5%, the *m* values are not consistent. Careful observation of the sintering strain curves revealed that at those temperatures where the *m* values were unreasonably large, the curves reached saturation and flattened at longer hold times. At those temperatures where the shrinkage did not saturate, the sintering exponents were estimated to be m_{1173}=1.46 and m_{1273}=2.14 (LD through defects and GB recreation respectively, in accordance with the models of Kingery et al, 1975 and Coble RL, 1958). This temperature range seems to be a transition regime between defect-assisted LD and the initiation of GB diffusion at higher temperatures. Irrespective of the sintering mechanism, the initial temperature range shows two characteristics: presence of a non densifying mechanism and end point densities.

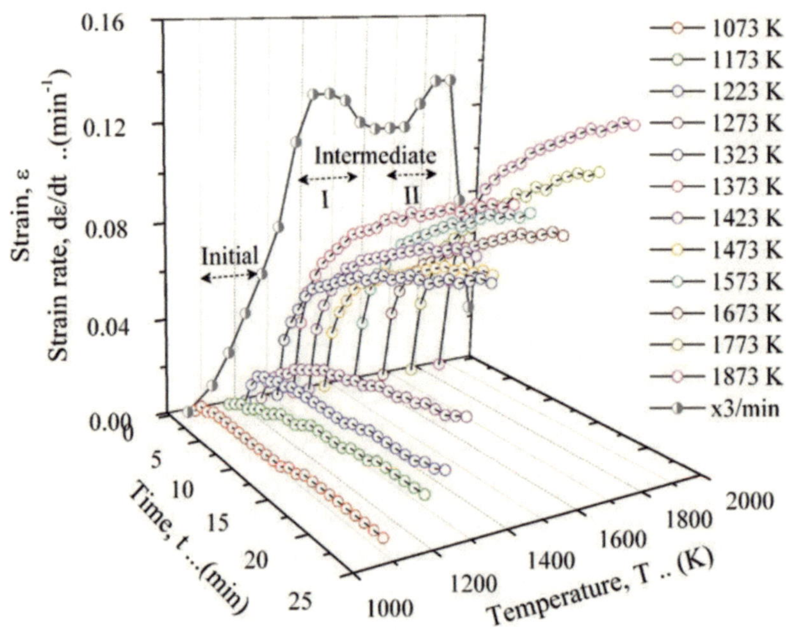

Figure 1. Isothermal and CRH sintering curves at different temperatures.

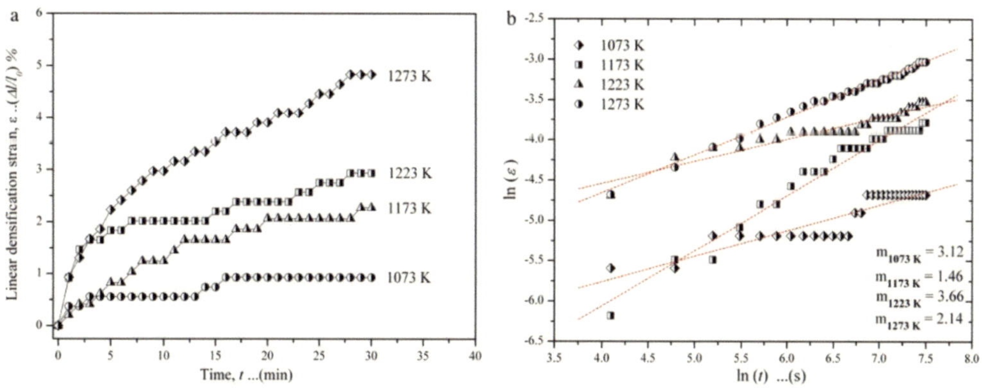

Figure 2. (a) Linear densification strains from 1073 K – 1273 K and (b) the corresponding sintering exponents calculated according to eqn. (4).

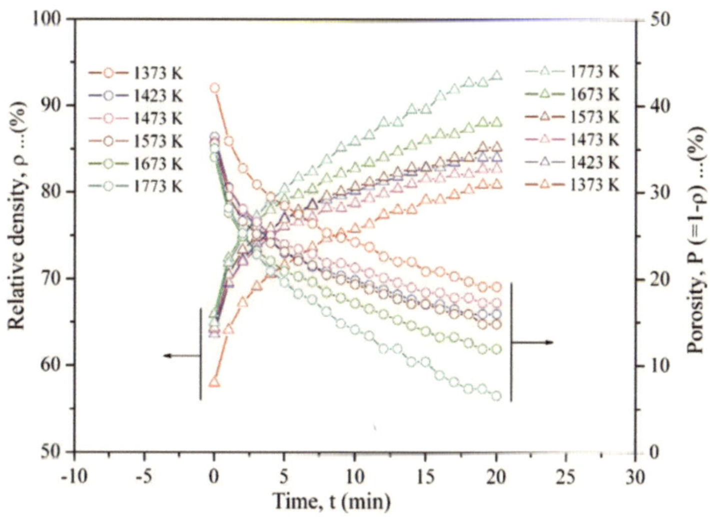

Figure 3. Porosity and relative densities at different intermediate temperatures.

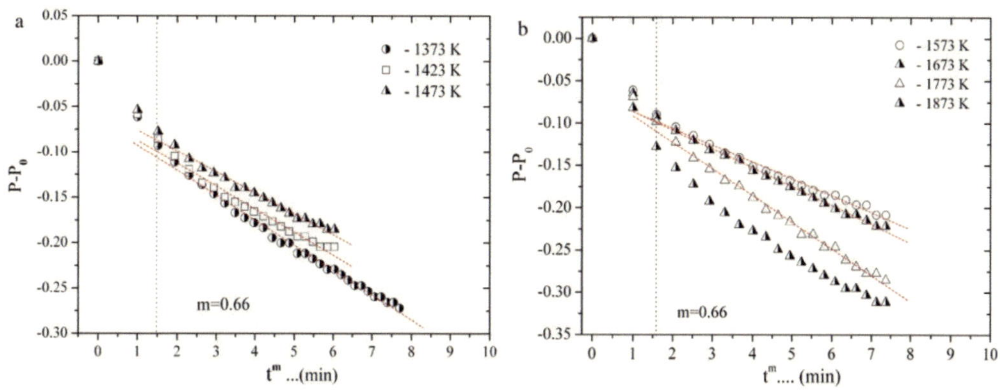

Figure 4. Plots of P-P_0 vs. t^m according to eqns. (7)-(9) between (a) 1373 – 1473 K and (b) 1573-1873 K.

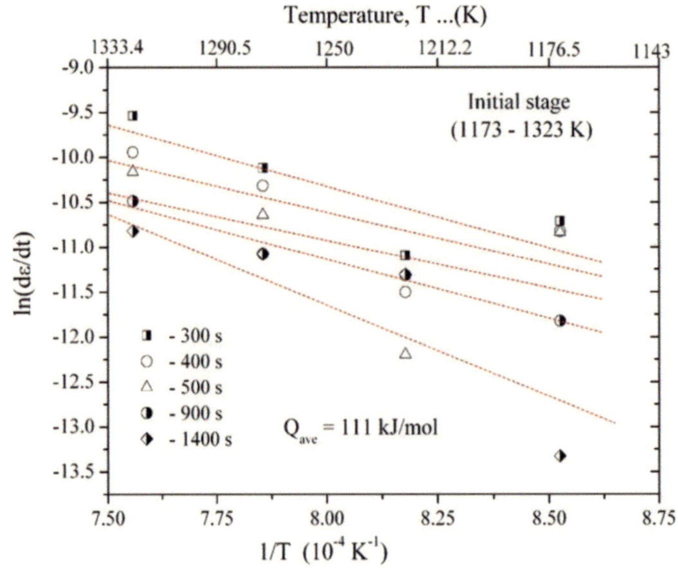

Figure 5. Calculation of apparent activation energy by Dorn's method.

For analysing the intermediate stage, the porosity fraction was estimated as $P = 1 - \rho$, where $\rho = (1 - \varepsilon)^3 \rho_f$ is the instantaneous density and ρ_f is the final density expressed as a fraction of the theoretical density. **Fig. 3** shows the porosity and relative densities of the samples at different temperatures in the intermediate stage. At the start of the isothermal hold period, the porosity was ≈35 to 42% (at various temperatures) which decreases to a value between 6 and 18% at the end of the hold period. It is interesting to note that although the density increases with the hold time, they are almost constant in a narrow range of temperature (1400 to 1573 K). The end density seems to be a strong function of the initial density at $t = 0$. **Fig. 4a,b** shows the subsequent kinetic analysis of the intermediate stage obtained by plotting P-P_0 against t^m. Most of the data points fall in a straight line when m=0.66, suggestive of Coble's grain boundary dominated sintering mechanism.

The apparent activation energy of sintering was calculated using the Dorn method. Only positive values of slope were considered. In the designated initial stage from 1173 K to 1323 K (**Fig. 5**), Q = 111 kJ/mol. In the final stages (1673– 1823 K), a small activation energy of 45 kJ/mol was calculated (figure not shown). The other temperature ranges could not be analyzed without ambiguity since sintering strains between the temperatures varied rapidly and our sampling interval (every 50 or 100 K) was inadequate to collect sufficient data points. The CRH experiments were hence considered for analysis at higher temperatures.

The sintering kinetics from the CRH experiments was also analysed. **Fig. 6** shows a plot of ln($Td\varepsilon/dT$) vs. $1/T$ along with the measured values of the effective activation energy. Low heating rates were found to show transition stages clearly. Three different sintering stages can be identified from 1173 K to 1873 K by the change in slope: a first stage ranging from 1173 to 1273 K with mQ = 56.7 kJ/mol, a second stage from 1323 to 1473 K and mQ = 103.5 kJ/mol and a third stage with mQ = 41.35 kJ/mol between 1673 and 1823 K. Consistent with the results of the Dorn method shown earlier, there was a narrow range with negative slope in the CRH experiments also between the second and third regions. The activation energy for sintering controlled by lattice diffusion ($m = 1/2$) in the I stage is Q I = 113.4 kJ/mol which agrees very well with the calculations of the Dorn method for isothermal sintering (Q = 111 kJ/mol). In the second stage, assuming GB diffusion ($m = 1/3$), Q I = 310.5 kJ/mol which closely corresponds to the activation energy for GB diffusion of C in WC [Bushmer C P and Crayton P H, 1971]. It should be mentioned however, that the appearance of this, 'second stage' depends on the heating rate (and consequently, the activation energy of the second stage is also a function of the heating rate). At low heating rates, a clear division between the first and second stages can be discerned by a change in slope, but at higher heating rates, it is impossible to differentiate between the first and second stage. The third stage clearly shows a very low activation energy, which could not be correlated to any reported solid state diffusion mechanism.

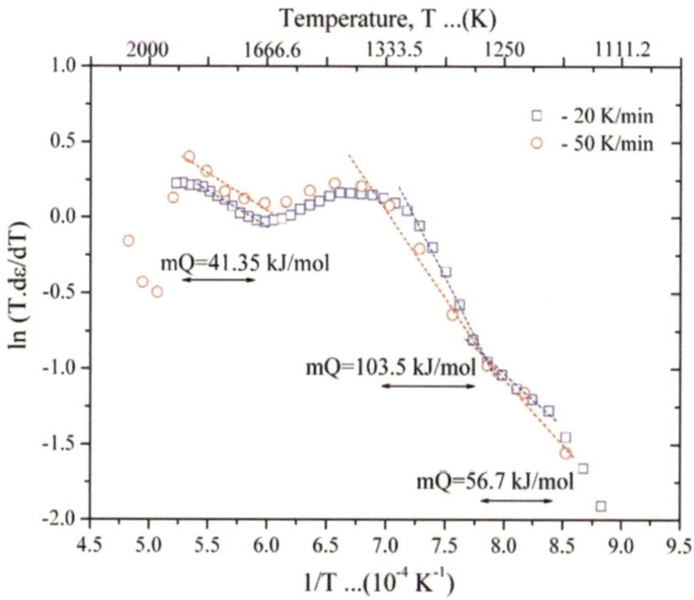

Figure 6. Calculation of effective activation energy from CRH experiments.

5.2. Microstructure analysis

A preliminary examination of the cross sections of the samples revealed that the edges of the completely densified compact was different from the bulk of the sample. **Fig. 7** shows the cross section SEM image and composition map of the sample by EPMA.

Clearly, huge abnormal grains populate the microstructure from the surface to a depth of nearly 30-40 μm. Interestingly, the chemical analysis of the surface by wavelength dispersive EPMA (Electron Probe Micro Analysis) also revealed a C deficient, W_2C layer on the surface. (It should be noted that the spatial resolution of the EPMA is rather low and therefore, while the W-rich layer on the surface is shown to be continuous, the region may actually comprise

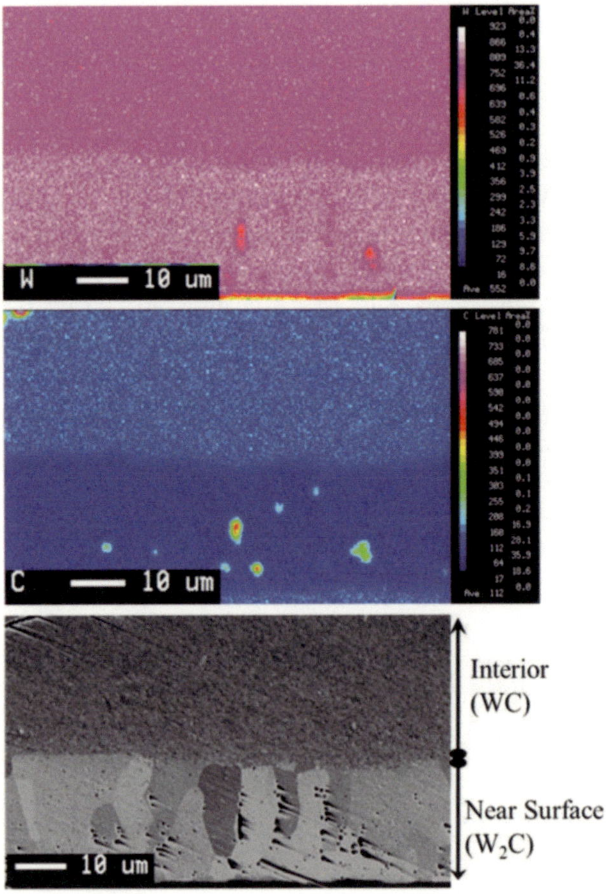

Figure 7. A cross-sectional composition map by EPMA near the graphite/WC interface of a completely sintered compact.

many small clusters of W_2C grains). Such differences in microstructure can occur by temperature gradients in the sample, resulting in a change in chemical composition at the punch/sample interface owing to the high activity of carbon in WC. Both hardness and fracture toughness measured on the surface and the interior showed that the surface was softer than the latter. With increasing heating rate, the grain size decreased with a corresponding increase in hardness, in accordance with the Hall-Petch effect, as reported elsewhere [Kumar A K N *et al*, 2010]. At higher loads, the hardness saturated to ≈2700 HV for the sample with the smallest grain size (with a sintering rate of 150 K/min, the final measured grain size was <300 nm), as shown in **Fig. 8**. The microstructure was also not uniform on the surface. The two phase regions existed as patches and were clearly discernible in the optical microscope. Indentation in these areas led to brittle fracture at the corners of the indent **(Fig. 9)**. Such a drastic change in the mechanical properties confirms the existence of W_2C, which is an embrittling phase in the W-C system [Luca Girardini *et al*, 2008]. More quantitative measurements of grain size and distribution were made using EBSD. The unique grain map (**Fig. 10a,b**) and quantitative grain size histogram plots measured from the area fraction of the grains (**Fig. 11**) showed a bimodal grain size distribution in the surface with the peaks at ≈700 nm and 1500 nm, while in the interior, the grain size distribution was also bimodal but with the two peaks at ≈250 nm and 480 nm. The bimodal size distribution arises because of abnormal grain growth (AGG) – a characteristic trait of the carbides that exhibit facetted grain boundaries [Li *et al*, 2007, Byung-Kwon Yoon *et al*, 2005]. It is also interesting to note that the average grain size of both the normal and abnormal grains is higher on the surface than in the interior.

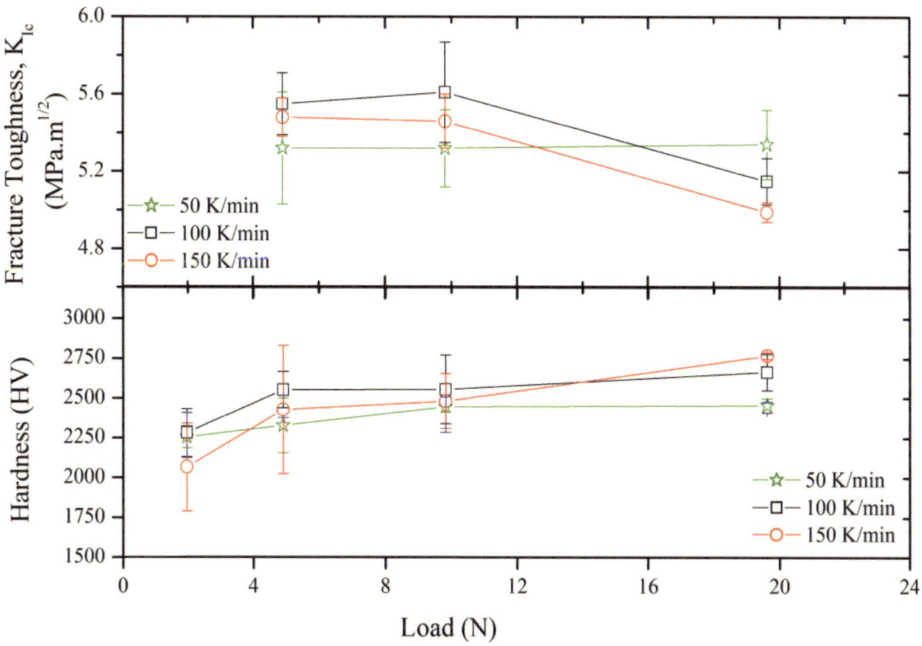

Figure 8. Hardness and fracture toughness of sintered *n*-WC compacts at different loads.

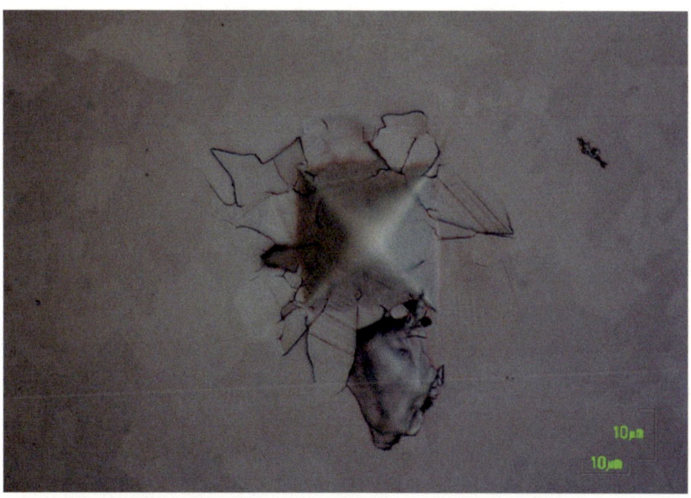

Figure 9. Brittle two-phase regions on the surface leading to indentation cracking.

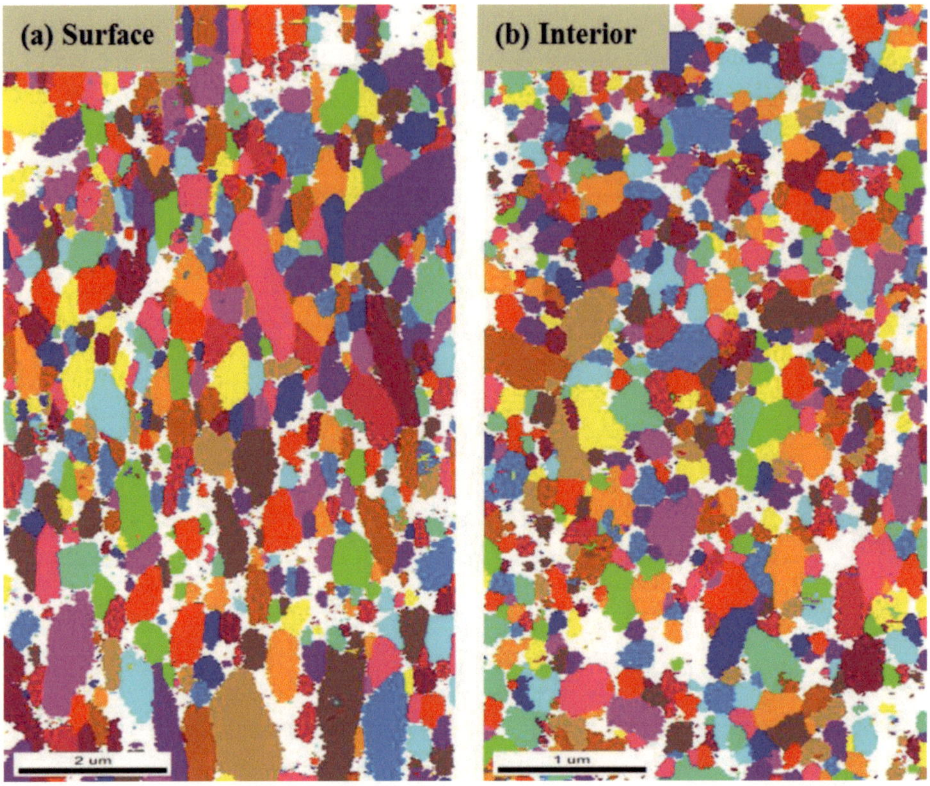

Figure 10. Unique colour grain map of the surface and interior of the samples by EBSD clearly showing larger grain size on the surface of the specimens.

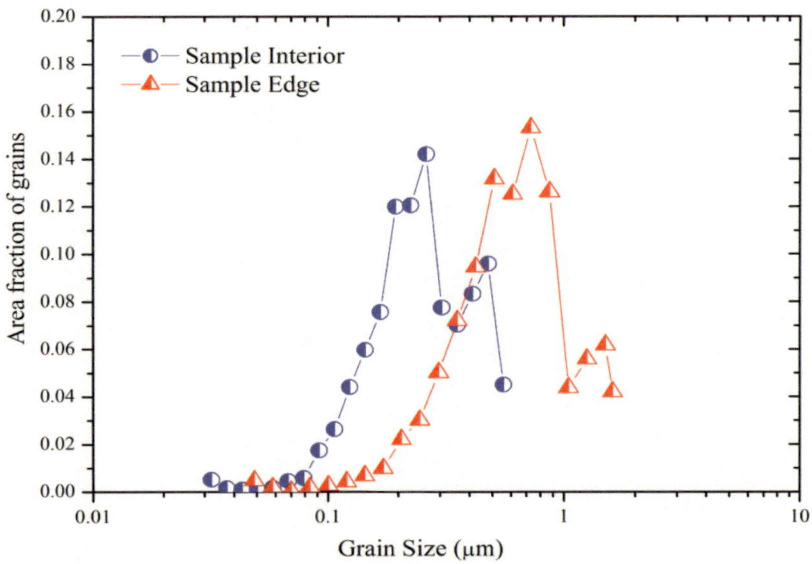

Figure 11. Grain size distributions by EBSD showing a bimodal distribution both on the surface and interior of the samples.

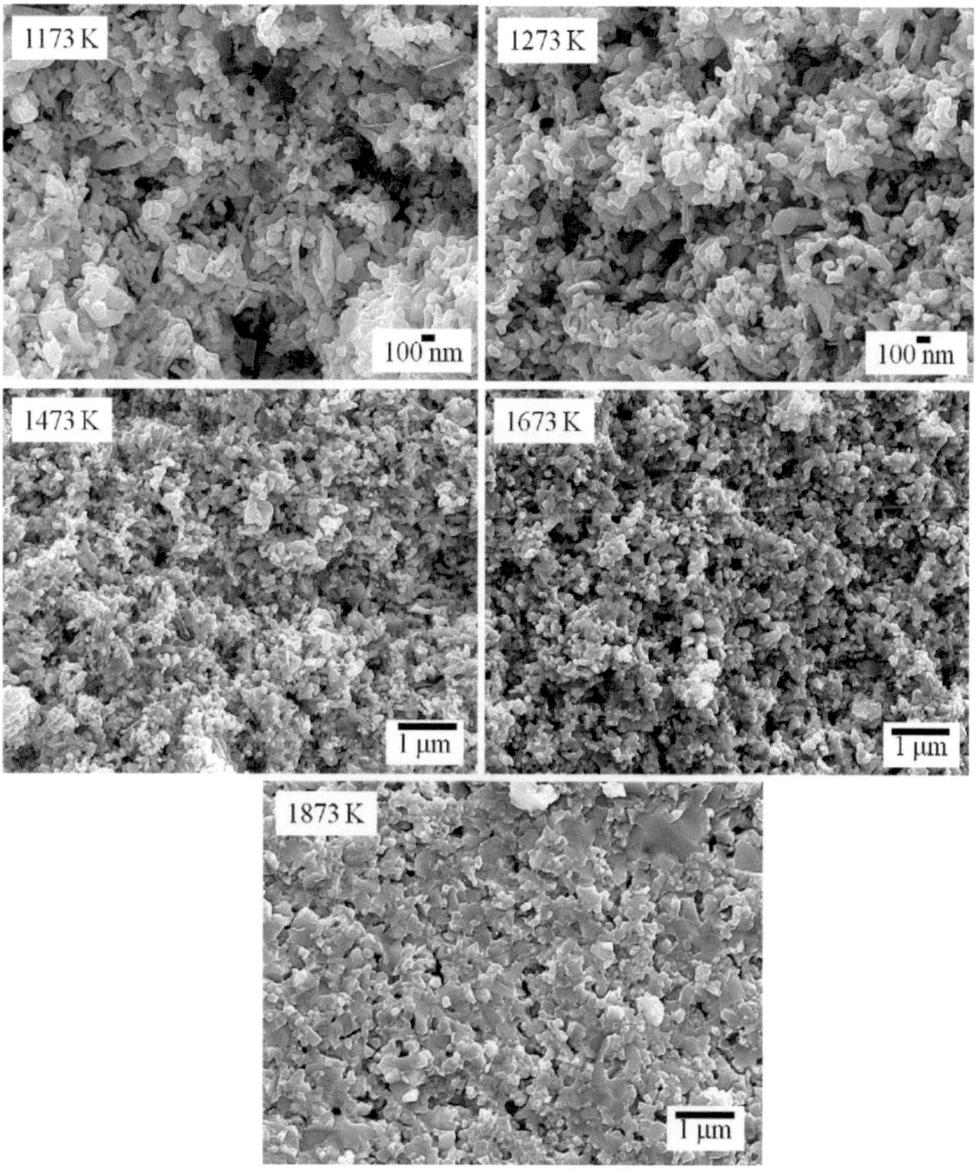

Figure 12. Microstructures of the compacts interrupted at various temperatures during sintering.

Fractured surfaces of the sintered compacts at various temperatures, observed by SEM are shown in **Fig. 12**. From 1173 K to 1323 K, the individual particles and bonded particles with necking can be discerned as a dispersed phase indicating the initial sintering stage. A few agglomerates can also be seen. From around 1373 K to 1773 K, large continuous pores were evident and this temperature range was considered to represent the intermediate stage of sintering. At 1873 K, most of the pore phase is pinched off, leading to the final sintering stage. However, the actual transition from the initial to intermediate stage sintering is rather vague as there are strong density gradients in the microstructure due to agglomeration. But as a preliminary estimate, micrographs from these temperature ranges combined with the CRH-Iso sinter curves mentioned earlier can be assumed to represent the different sintering stages. While the SEM analysis does reveal the formation of agglomerates – thereby partly explaining the humps and dip in the CRH curve - the observation still does not account for the low activation energies measured by the kinetic analysis.

To probe the structure of the sintering particles further, the interrupted samples were also observed by TEM. A few samples were selected to understand the sintering behaviour: the original WC powder, samples sintered to 1323 K, 1473 K, 1673 K and the final densified compact. The WC powder was simply put on a grid and observed. **Fig. 13** shows a few micrographs of the powder sample viewed under the TEM. It was surely not a mono disperse powder. Agglomeration was clearly obvious and interestingly, a substantial fraction of particles containing stacking faults (SFs) were also seen. The extensive streaking of the spot patterns confirm that the steps observed on the particles are indeed SFs. The faults extended right across several grains diametrically to a length of nearly 2X the particle size, resembling shear bands. As no mechanical milling was conducted, it is likely that the SFs were introduced into the particles during the production stage itself. The clear proof of the occurrence of SFs in the initial particles is an important observation since lattice defects can impact the activation energy for diffusional sintering. Diffraction studies also revealed that the SFs were present only on the prismatic$\{10\bar{1}0\}$ planes and the basal $\{0001\}$ planes were relatively free of defects.

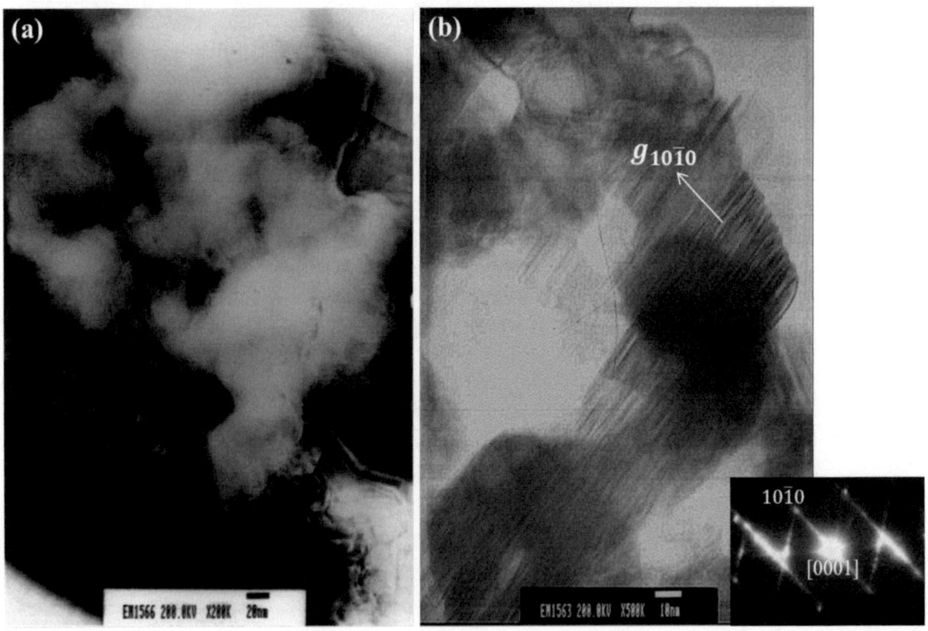

Figure 13. TEM micrographs of the *n*-WC powder showing agglomeration and stacking faults.

The sample sintered at 1323 K was observed next. It clearly showed signs of undergoing initial stage sintering (necking) in some of the separate particles that could be observed. The necks between particles were almost 5 nm thick and interestingly, the neck and the entire surface of most of the particles showed a sort of spotty, recrystallized-like phase. This phase was marked by its characteristic dull appearance and hardly showed any diffraction contrast. While particle re-deposition during PIPS is most probably the reason for this observation of an amorphous surface layer, in a later section we also consider the effect of local temperature gradients leading to surface overheating of the powders that can be expected in SPS. There was a high density of thin SFs on the prism planes in these samples too (**Fig. 14a-c**).

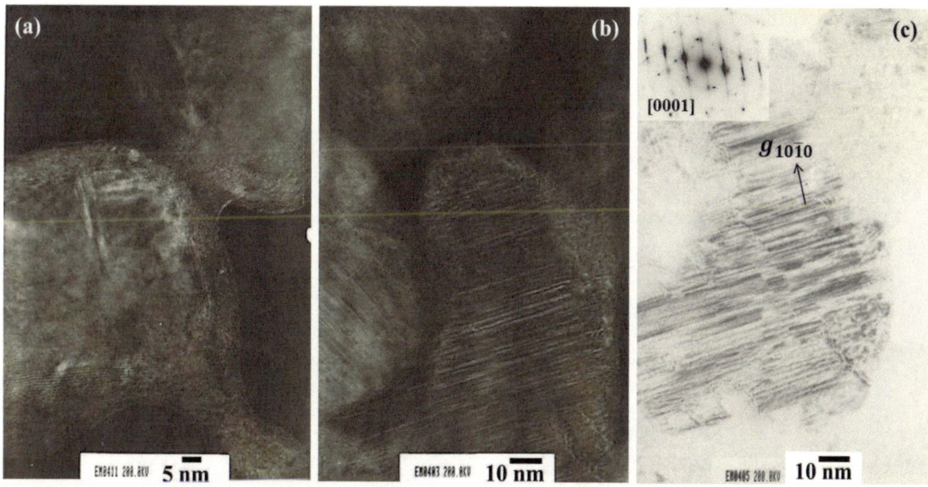

Figure 14. TEM micrographs of a sample sintered to 1323 K showing (a) necking (b) SFs with a thin amorphous GB phase and (c) Diffraction pattern (DP) and a dark field (DF) image confirming that the SFs populate the prismatic planes only.

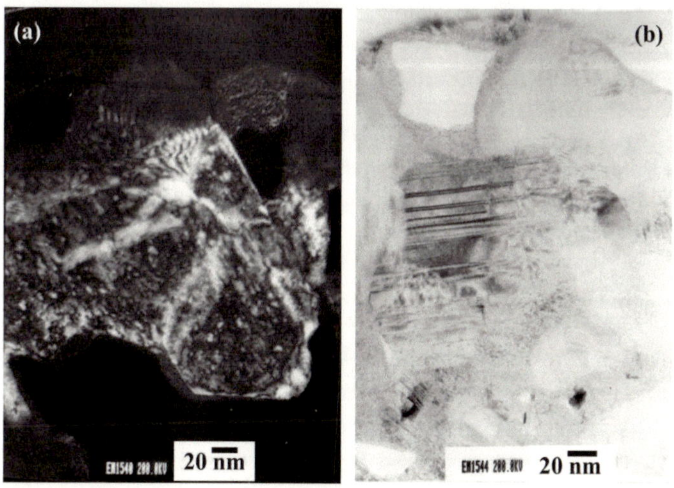

Figure 15. Sample sintered at 1473 K showing (a) a large grain with 3 or more sub grains and (b) SFs on prism planes.

The sample sintered at 1473 K showed evidence of necking, agglomeration and slight grain growth. In addition, a number of SFs could also be detected on similar prismatic planes – a continuation of the feature observed in the powders and the previous sample (**Fig. 15a-d**). The faults were well-formed and the fault line density in the observed grains was found to be lesser than that in the original powders.

The fourth sample that was investigated (1673 K) also showed the same features as that of the earlier sample sintered at 1473 K (**Fig. 16**). Necking was not observed, while the SFs were rather few and the grains were more facetted and clearly visible. In essence, the features were quite similar to the previous sample, except for a slight variation in the fraction of the phases and size of the grains. This sample also appeared to be in the intermediate stage of sintering.

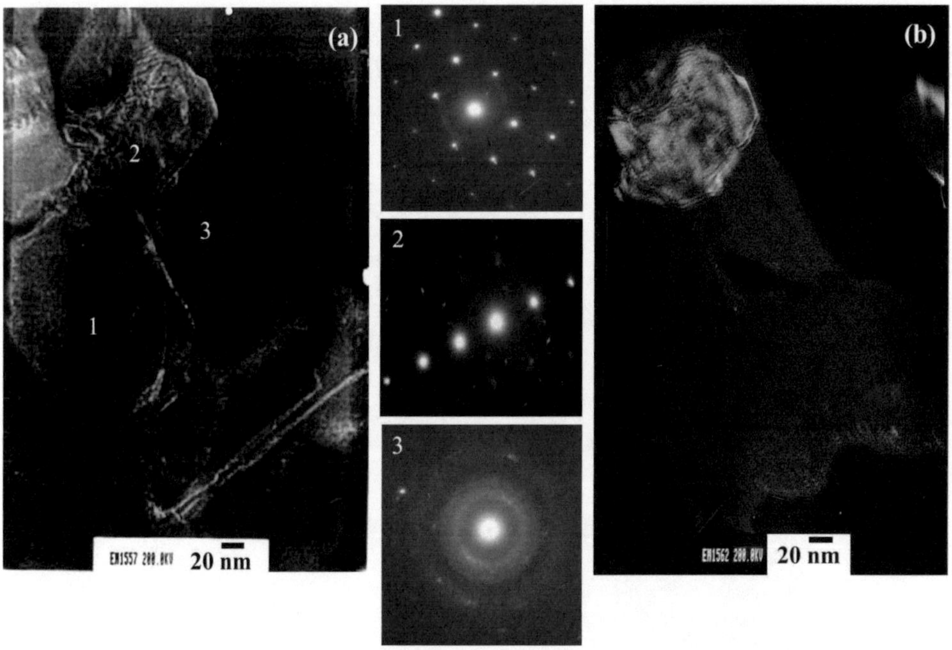

Figure 16. Sample sintered at 1673 K showing (a) three regions marked 1, 2 and 3 and their corresponding DPs. 1 is an almost defect-free grain imaged along $[1\bar{2}10]$, 2 contains GB dislocations as seen from the multi beam condition in the corresponding DP and 3 is an amorphous pocket with a diffuse ring pattern (b) is the DF image from an excited spot in the DP in 2.

The final sintered sample (2073 K) showed well-formed grains (**Fig. 17**). While the specimen still contained some SFs in the small grains, in some of the larger grains instead of the SFs, twins were also observed (confirmed from the DPs which showed twin reflections). Interestingly, small grains of the semi-carbide W_2C measuring ≈50-100 nm could be seen in the sample (TEM samples were prepared from the cross section and not surface). All the grains were faceted and had sharp GBs. The grain growth into such well-formed structure seems to occur rather rapidly in the final stages of sintering with the annihilation of SFs, removal of the amorphous pockets and pore closure.

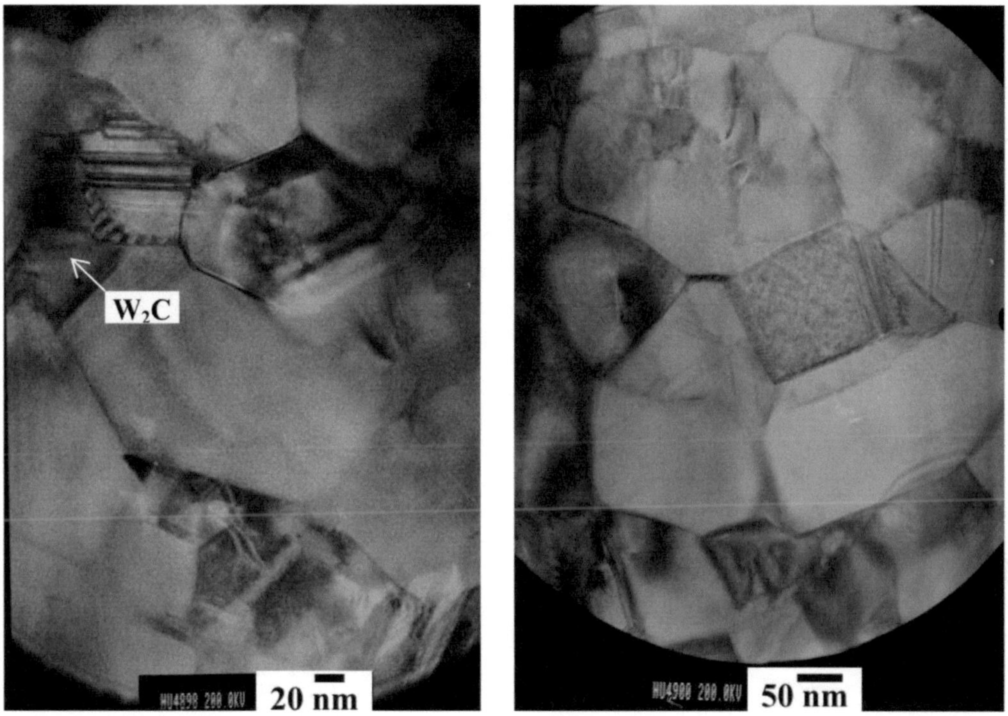

Figure 17. Sample sintered to almost full density at 2073 K showing a few SFs, but mostly well- formed faceted grains.

6. Discussion

The significant results of the kinetic and microstructural analyses detailed earlier are presented in an integrated way in **Fig. 18**. The measured relative densities and the corresponding grain size evolution together represent the sintering trajectory of the *n*-WC powder. It is seen that densification dominates during the initial stages up to ≈ 1350 K with the relative density increasing from ≈ 68% to 85%. Following this rapid densification, the density then *decreases* slightly; interestingly, grain growth is also insignificant during this stage. This is a surprising observation since while the densification rate ($d\varepsilon/dt$) can decrease owing to many factors (grain growth or the formation of a reaction product etc.), an actual decrease in the measured density cannot occur *unless the compact is subjected to a volumetric expansion or unless there is pore growth in the specimen*. A more clear picture arises when we convert the isothermal strains at different temperatures shown earlier in **Fig. 1** to instantaneous densities as shown in **Fig. 19**. We note immediately that near the vicinity of the first peak in the CRH experiment, while the isothermal densification strains are high, the *initial densities are also lower*. This simply means that the densification rate at any temperature is a function of the green density at that temperature. This behaviour persists over a small temperature range of ≈150 K after which at around 1500 K, the second stage of densification again begins; however, simultaneous grain growth is also observed here. This multi-stage sintering process can be explained only by a combination of different densifying and non-densifying processes occurring simultaneously and sequentially at different stages. Since the sintering behaviour (*viz.*, the sintering kinetics and microstructure evolution) is influenced by the nano size of the particles, nature of the powder (agglomerates) and the activated sintering process – the effects of any of which on either the kinetics or microstructure are not exactly known, - we split the discussion into different segments in order not to mix up the issues and hence lend more clarity and focus to our overall analysis. The chief aims of this section are therefore to interpret the following observed phenomena: (i) the low sintering activation energies, (ii) occurrence of multiple sintering stages and (iii) the decrease in measured density at intermediate temperatures.

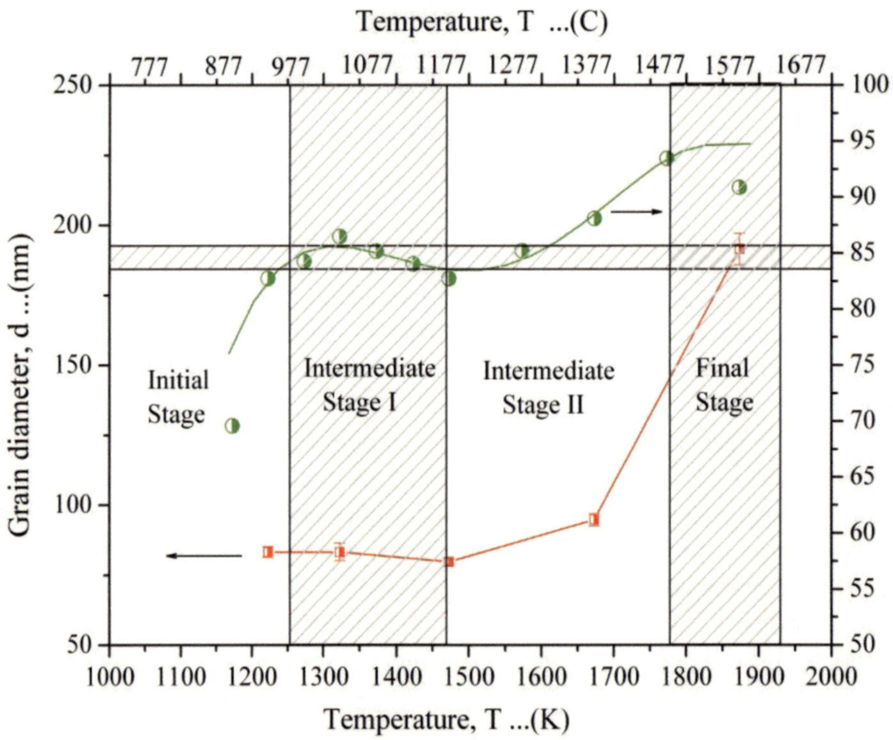

Figure 18. A sintering map showing variation of grain size and relative density with temperature.

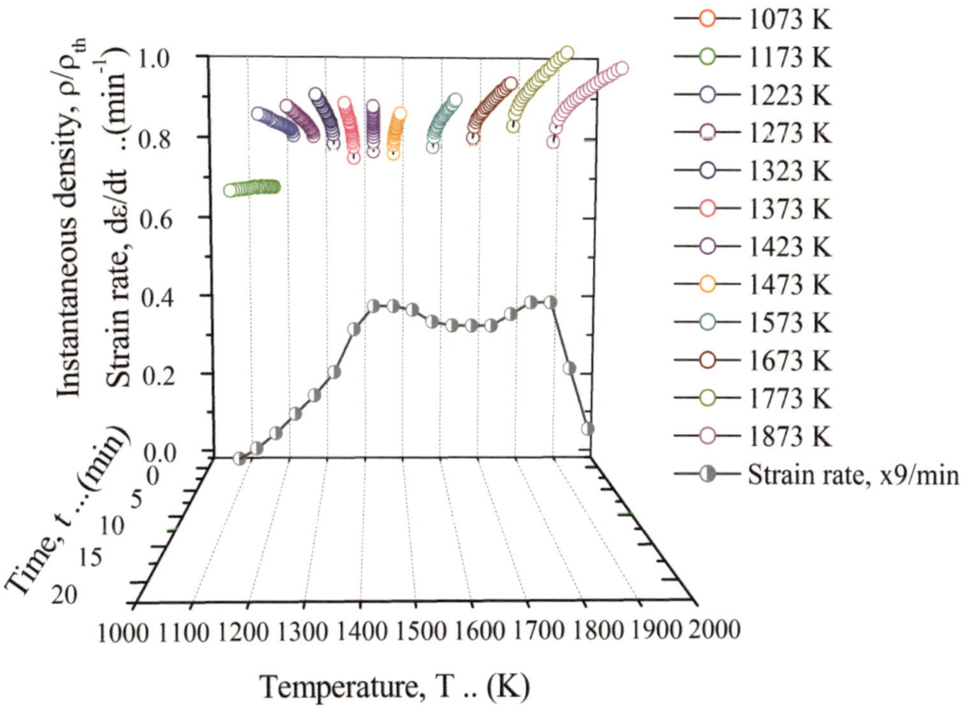

Figure 19. Comparison of CRH sintering rate and instantaneous densities during isothermal sintering (calculated by measuring the final density and using the relation, $\rho_i=(1-\varepsilon)^3\rho_f$)

6.1. Size effect on the sintering kinetics

Sintering of compounds (both ionic and covalent) occurs by the slowest species diffusing through the fastest route to establish the chemical equilibrium associated with the stoichiometry of the compound. In WC, carbon is generally considered to be more mobile than W. The crystal structure of WC is hcp (c/a=0.985) with C atoms occupying the interior positions in the unit cell. Yet, carbon cannot be equated to a regular interstitial atom in WC. This is because the tendency for formation of the W-C bond is quite strong. The origin of defects and the lower mechanical strength along the prism planes rather than on the basal planes can be traced back partly to this molecular origin [Nabarro F R N *et al*, 2008]. Yet, many direct and indirect evidences are available to support the fact that at higher temperatures, C is extremely active than W. For instance, the diffusion of tracer Carbon ($_{14}$C) in WC was found to occur initially by LD and later on by GB diffusion through the WC grains [Bushmer and Crayton, 1971]. Estimated values of D_v was very low compared to GB diffusion with the diffusivity ratio, D_{gb}/D_v being of the order of 10^3 (D_{gb} ~ 10^{-9} m^2 /s at 2238 K). The activation energy for GB diffusion of $_{14}$C was estimated to be ≈ 300 kJ/mol. Another estimate provides a value of 10^{-22} m^2/s for lattice diffusion of $_{14}$C in WC [Andrievky R A and Spivak I I, 1983]. Activation energies for the diffusion of W can be assumed to be higher. It should be noted that these estimates can be different when a liquid binder like Co is present since the WC molecule has to first dissociate into W and C, dissolve in the liquid and then migrate to the re-deposition surface. Therefore in solid state sintering without Co, it might be expected that the diffusivity of the relatively immobile W should control the effective diffusivity during the sintering of WC. Yet surprisingly, our analysis suggests that the densification processes are activated by energies as low as 100 kJ/mol. There is a substantial body of evidence that suggests similar values of low activation energies that are comparable to our present results during the initial stage sintering of either WC nano particles or by activated sintering of large WC particles (discussed in the following section). Mugenstein and co-workers [Goren-Mugenstein G R *et al*, 1998] studied the initial stage sintering of WC powders with an average particle size 100-500 nm by conventional furnace sintering and employing the Dorn method, determined the activation energy to be 76 kJ/mol. Our calculations described in the earlier sections too showed values to be of the same order. Fang and co-workers [Fang Z *et al*, 2004] studied the sintering of various grades of WC-Co in a vacuum furnace. While they did not measure the kinetics, they did indeed observe that the onset temperature of sintering of the nano sized powders (50 nm particles) started 160 K below that of the micron sized powders. Moreover, since the nature of the sintering curves was similar for both particle sizes, they surmised that the densification steps were similar irrespective of particle size, but that the activation energy decreased with the size of the particles. Therefore, it seems that lowering of the activation energy can be achieved merely by reducing the particle size, atleast during the initial stage. Studies on certain other nano sized oxide ceramics also support the same view. Theunissen and co-workers [Theunissen G S A M. *et al*, 1993] studied conventional sintering of chemically synthesized ultrafine (8-50 nm) Y_2O_3-ZnO_2 ceramics and found that the activation energy was as low as 100 kJ/mol, which again did not correspond to any densifying diffusion mechanism. Comparison of these and many other results available in the literature [see for example, Dominguez O and Bigot J, 1995 (*n*-Fe), Kinemuchi Y and Watari K, 2008 (*n*-CeO_2), Victor Zamora *et al*, 2012 (*n*-ZrB_2), Li J G and Sun X, 2000 (*n*-Al_2O_3)] strongly support the view that a mere reduction in particle size can lower the activation energy for sintering. Fundamentally, this relates to a scaling down of a thermodynamic quantity with particle size, which can be analysed using Herring's scaling laws [Rahaman M N, (2003), Wenming Zeng *et al*, (1999)]. These laws basically compare the rate of sintering (densifying and non densifying) in different pathways for two dimensionally different particle systems. **Fig. 20** shows the sintering rates for two systems 1 and 2, as a function of the particle size ratio (R1/R2). The graph shows a cross-over when the particle size decreases (GB and Surface diffusion are significantly enhanced when R1<<R2). All the possible sintering mechanisms in a system can therefore be weighed on this scale simply as a function of particle size, provided all other factors are unchanged. Zeng and co-workers [Wenming Zeng *et al*, (1999)] used a similar type of semi-quantitative analysis and pointed out that, theoretically the sintering temperature (and hence activation energy) of pure α-Al_2O_3 can decrease from 1773 K for an initial powder size of 600 nm to 1498 K for LD and even lower to 1423 K for GB diffusion as the particle size decreases to 60 nm. In support of their argument, they even point out to some relevant experimental reports published elsewhere. The most probable explanation for all these various observations is that the surface and GBs of nano powders are easily activated and play a major role in lowering the sintering temperature and activation energy, irrespective

of whether other diffusion routes (LD, diffusion through defects etc.) dominate sintering in similar systems with a larger initial particle size. The same phenomenon can also be assumed to occur in *n*-WC powders. The presence of excessive planar defects also obfuscates the atomic diffusivity leading to rapid sintering. From the TEM micrographs, it is seen that the SFs dominate the microstructure until 1473 K, which act as short-circuit diffusion paths for sintering. However, while the particle size can be argued to lower the activation energy, it still does not explain the end point densities observed in most of the initial temperature range (shown in **Figs. 1, 19** earlier). Neither does the fact that surface diffusion is enhanced imply that shrinkage is also enhanced, since surface diffusion, unlike GB diffusion, is a non densifying sintering mechanism. Therefore, in addition to the particle size effect, there are other factors that control the sintering of the *n*-WC powders in the present case.

Direct observation of the presence of hard agglomerates (described in the following section) and the flattening of the shrinkage strain curves at low temperatures point to a mechanism of *particle rearrangement* (PR) that can induce a rapid initial densification but eventually leads to a saturation density. PR usually occurs at low temperatures, starts with a minor shrinkage and leads to an end point density when the closest packing is reached. This type of densification by rearrangement can be enhanced in the presence of surface diffusion, and GB sliding – both of which can be assumed to occur in the green compact. Most of the defects observed in the powders and low temperature compacts run diametrically across the particles and therefore can be equated to GB dislocations which can easily lead to GB sliding. The low sintering activation energy observed during the initial stage may therefore be attributed to a densification mechanism brought about by a combination of PR assisted by surface diffusion (SD) leading to GB sliding. This mechanism explains not only the low activation energy but also the saturation shrinkage strains observed in the low temperature region of the Iso-CRH curves.

6.2. Influence of agglomeration

Nano powders, owing to their high surface area to volume ratio, are characterized by a high surface energy. This leads to a difference in the chemical potential of the atomic species constituting the particle at the interior and surface and forms the chief driving force for agglomeration or aggregation. Such agglomerated nano powders are characterised by small groups of particles demarcated by GBs that in turn coalesce to form larger aggregates with pore boundaries [Lange F F, 1984]. This results in a totally non uniform microstructure leading to differential densification and multiple routes to sintering. Hence, the concept of the fastest diffusion route during sintering becomes complicated as intra agglomerate pores may densify easily while the larger pores may require higher energies for densification.

Fig. 21 shows a high magnification FE-SEM micrograph of a compact interrupted at 1073 K and also the initial powder, which shows a composite phase consisting of both individual particles and clusters of connected particles that have undergone necking. The clusters are hard agglomerates that persist even after the application of external pressure (40-50 MPa).

Unlike the soft agglomerates that form by weak van der Waals/electrostatic bonding and constitute inter agglomerate bridges, the hard agglomerates are formed by solid state diffusional bonding. In all the samples studied, the green density (before sintering) was less than 43%. It is clear that while the initial stage may be controlled by PR, the intermediate stage is governed by agglomerate evolution. When agglomerates form, internal density gradients are set up leading to a large pore size distribution. Consequently, sintering sub-stages are introduced in the intermediate stage by the differences in the sintering kinetics of the inter and intra agglomerate pores.

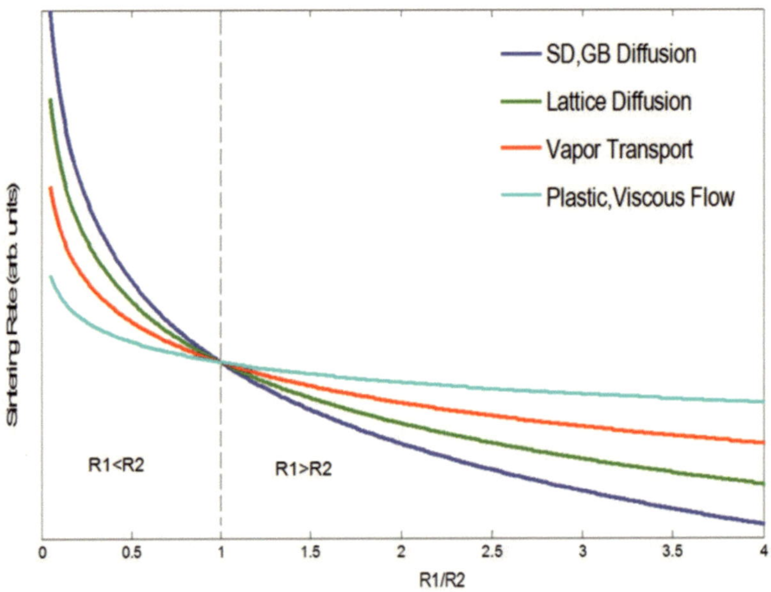

Figure 20. Dominance of various sintering mechanisms as a function of particle size calculated using Herring's scaling law.

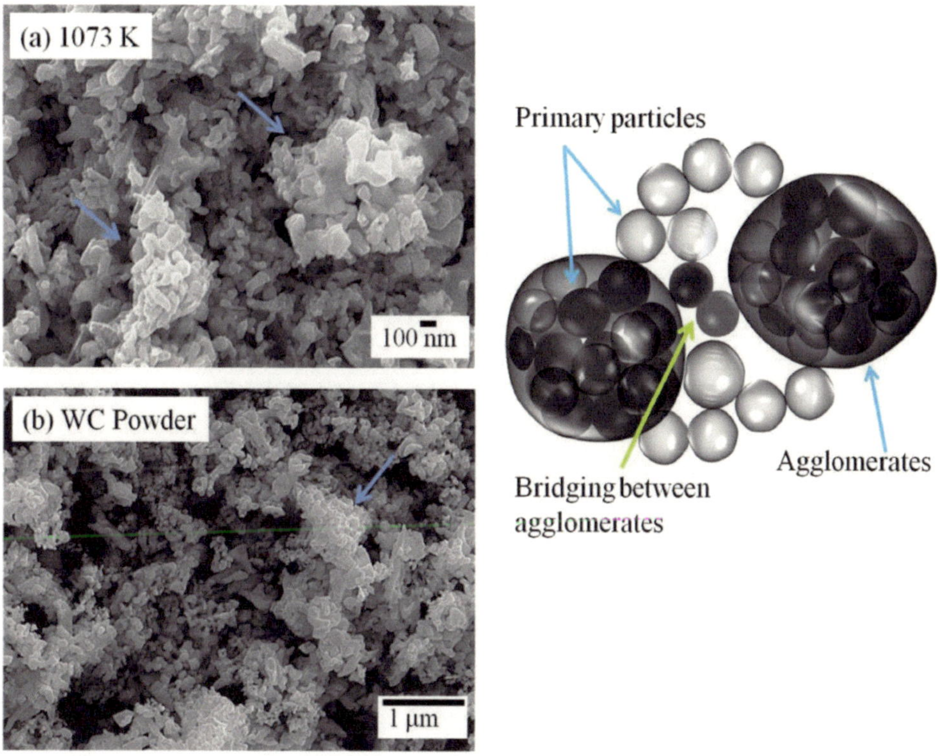

Figure 21. FE-SEM image of WC powder and compact sintered at 1073 K showing agglomeration. On the right is a schematic of the general low temperature microstructure.

Fig. 22(a–c) shows the high magnification microstructure of the samples from 1373 K to 1573 K. The agglomerates are enhanced and importantly, two different types of pore morphologies can be clearly distinguished: long, continuous inter-agglomerate pores and small disconnected intra-agglomerate pores (inter-particle pores). With increase in temperature, the individual agglomerates densify by sintering and slight grain growth, while there is not much observable change in the nature of the inter-agglomerate pores. At 1673 K, the grains can be clearly discerned and the intra-agglomerate pores have almost vanished, replaced by continuous pores (**Fig. 22d**). At still higher temperatures, (1873 K), the continuous pores become isolated and pinched-off resembling the final sintering stage (**Fig. 22e**). It is clear that the intermediate stage and much of the entire densification process is governed by agglomerate evolution.

Presence of hard agglomerates can partly explain the occurrence of sub-stages observed in the sintering rate curves. Initially, at low temperatures (T≤1423 K), there is a rapid increase in the densification rate of the compact. This occurs both by compaction of the agglomerates (contribution of intra-agglomerate sintering, which is expected to be low) and by rearrangement of agglomerates (inter-agglomerate sintering). The end densities increase to around 80%. This initial rapid shrinkage is followed by a saturation of the densification rate in the CRH curve. But interestingly, isothermal holds at these temperatures seem to induce high sintering strains. From this until ≈1623 K, the sintering rate decreases while the

Figure 22. High magnification FE-SEM images of the intermediate sintering stage showing evolution of agglomerates.

isothermal sintering strains continuously increase. The decrease in the non-isothermal shrinkage rate can be explained on the basis of the energetics of the sintering of agglomerated powders: following the formation of stable agglomerates, the fraction of intra agglomerate pores is significantly reduced. From this point, the sinterability of inter agglomerate pores controls the densification rate. But this is also prevented because of the large pore co-ordination number N_c. The dihedral angle (φ) in WC– Co is reported to vary between 30° and 120° [Gurland J, 1977]. WC (and in general, most of the transition metal carbides) exhibit facetted GBs and therefore the dihedral angle can vary drastically between different GBs. However, a statistical average of 75° can be assumed for φ. The approximate value for the critical pore co-ordination number, N_c can then be calculated from [Kingery et al, 1975]:

$$N_c = \frac{360}{180-\varphi} \tag{10}$$

where φ is the dihedral angle between the particles in the agglomerate. It turns out that N_c in this case is 3.6. Our microstructural observation show agglomerates that are co-ordinated by a far larger number of particles (**Fig. 22**) which clearly explains why the agglomerate sintering rate faces a thermodynamic barrier. This inter-agglomerate pore stability retards the shrinkage rate in the intermediate stage. Following the dip in shrinkage strain rate, there is a passive period over which the system tries to evolve by *particle growth within the agglomerates*. The massive grain growth leads to a breakup of the agglomerate identity into large grains and the stable inter agglomerate pores now start to sinter. As the agglomerates continuously convert into large grains, the pores start to shrink rapidly, surpassing the grain growth rate. Hence concurrent grain growth supports shrinkage until the continuous pores are eliminated and the isolated pore structure resembling the final stage appears. The relative densities increase to nearly 94% during this stage. The final stage of sintering is reached at temperatures above 1773 K when the pore phases are pinched off which again results in a reduction of the densification rate. Bimodal shrinkage rates seem to be a signature trait of agglomeration-induced densification and have been commonly reported in the literature [Lanfredi S *et al*, 2000, Jiang X X, 1994]. Non agglomerated powders show only one sintering rate maxima while agglomerated powders usually show double maxima for the bimodal pore size distributions [Nobre M A L *et al*, 1996, Shi J L *et al*, 1994, Duran P *et al*, 1996, Knorr P *et al*, 2000].

However, while the agglomerate evolution mechanism explains the sintering rate curve, it does not explain the instantaneous densification curves derived from the isothermal shrinkage data. As mentioned previously, the observation that the measured densities decrease with increase in temperature implies pore growth. Sometimes, coarsening can also lead to a decrease in density. But actual measurements do not show significant coarsening to occur in this temperature interval. Pore growth occurs to reduce the total free energy of the powder system:

$$\Delta G = \gamma_{gb}\Delta A_{gb} + \gamma_{s-v}\Delta A_{s-v} \tag{11}$$

where, γ refers to the energy of either a GB or pore and A_{gb}, A_{s-v} are the corresponding areas. Under conditions, when $\gamma_{s-v} < \gamma_{gb}$, pore growth can occur. This can happen particularly when there is oxidation of the particles at high temperatures that reduces the surface energy of the particles. X-ray diffractograms of the samples at various temperatures are shown in **Fig 23**. At about 1300 K, it is observed that small peaks of WO_3 show up. In addition, from the earlier TEM micrographs, it is clear that there is a tendency for dissociation of WC leading to oxidation of the particles. Therefore, it is probable that the oxidation of the WC particles could have led to the growth of pores. WO_3 has a high vapour pressure and therefore evaporates as it forms. It is probable that this continuous formation and evaporation of WO_3 leads to an increase in porosity (which lowers the density) while isothermal holding at these temperatures leads to an increase in densification due to the applied stress and continuous repacking of agglomerates. Liu *et al* [Dean-Mo Liu *et al*, 1999] reported a detailed study on the influence of agglomeration of zirconia (ZrO_2) powder. Their observation corresponds very well with our experiments: systems with a lower green density and which consequently are highly agglomerated show the maximum sintering rate and reach full density. The densification rate below the first maximum in the CRH curve corresponds to regions where isothermal sintering can be carried out. Similar results have been reported for *n*-MgO too by Itatani *et al*, [Itatani K *et al*, 1993] who show that a lower green density increases final densities of compacts. While these previous reports have not studied the isothermal and non-isothermal sintering behaviour at any particular temperature, they assume that coarsening could be the reason for the low densities at certain temperatures. This point is still ambiguous and requires further detailed investigations to clarify the actual mechanism. But as Kellet and Lange have pointed out earlier, for a fixed sintering temperature and time, the end point density is proportional to the bulk density of the powder [Kellet B J and Lange F F, 1983].

6.3. Influence of pulsed electric current

While the preceding discussions on particle size and agglomeration explains the multi-step sintering and partly explains the lowering of the activation energy, the effect of an electric field and high currents during sintering and their implications on sintering are discussed in this section. A thin recrystallized region between the WC particles can result by overheating at the neck regions – a characteristic of the SPS method. In the actual experiment, the external current flowing through the sample was found to increase continuously as the compact densified. The small particle size and the high current (~ 700 A at peak densification) can be expected to induce very high current densities on the particle surface. An approximate calculation of the local temperature gradient between the interior and surface of a nano particle can be carried out using a recent model of SPS proposed by Olevsky and Froyen [Eugene Olevsky and Froyen, 2009]. In their model of heat conduction in SPS, the local temperature gradient, without considering heat loss is given by:

$$\nabla T = \frac{1}{G+r_p}\sqrt{\frac{E^2 \Delta t \lambda_e T_0}{2nC}} \tag{12}$$

where, G is the grain size, r_p is the pore radius, C is the specific heat capacity, T_0 is the temperature from which sintering is assumed to start (873 K, in this case), E is the electric field (V/m), λ_e is the electrical conductivity, Δt is the total (ON+OFF) pulse sequence duration and n is the number of pulses required to reach the desired temperature. When applied to oxide ceramics with low conductivity like Al_2O_3, local temperature gradients of the order of 10^6 K/m were determined during SPS [Eugene Olevsky and Froyen, 2009].. In our experiments, we used an ON/OFF pulse ratio of 12/2 which corresponds to 39.6 ms ON time and 6.6 ms OFF time. The ON pulse comprises twelve 3.3 ms pulses. Other approximate values were also plugged in: $G \approx 100$ nm, $r_p \approx 50$ nm, $\Delta t = 46.2$ ms, $n = 120220$, $E \approx 1000$ V/m. The values of C and λ_e were obtained from the literature: $C_{1300K} \approx 0.0175$ J/K/m^3 [Andon et al, 1975]. For a sample with residual porosity P, the heat capacity is given by $C = (1-P)C_{1300K}$ [Eugene Olevsky and Froyen, 2009, Yann Aman *et al*, 2011]. Since data on the electrical resistivity of WC at high temperatures were unavailable, with a knowledge of the room temperature resistivity of WC ($\rho_{300K} \approx 20$ μΩ·m) and the temperature coefficient of resistivity $\alpha \approx 4500$ /K [Grebenkina and Denbnovetskaya, 1968], a linear approximation was applied from 300 K to the temperature of interest (1300 K) using the relation $\rho = \rho_{300K}(1+\alpha\Delta T)$. This approximation, although slightly in error, may yield a variation of one order of magnitude in the final result. The electrical conductivity ($\lambda_e = 1/\rho$) of a sample with residual porosity, P was calculated as $\lambda_e(P,T) = \lambda_e(0,T)[(1-P)/(1+2P)]$ [Eugene Olevsky and Froyen, 2009]. Using these values, $\nabla T \approx 60 \times 10^6$ K/m. Owing to these large gradients, a plasma is more likely to form at the neck area. However, the temperature can be rapidly conducted over the particle surfaces leading to surface melting in a zone of 5-10 nm at the periphery of the grains as observed near the neck of the sintered particles. These regions can then enhance sintering by reducing inter-granular friction leading to more compact packing (densification by particle rearrangement) and provide easy routes for GB diffusion. At higher temperatures or in densely compacted regions, the term, 'grains' is more appropriate than, 'particles'. It was observed that such regions where the microstructure can be described as 'grains', are not surrounded by the recrystallized phase. While the actual observation of a GB phase during SPS of WC has not been reported yet, certain recent investigations by Demiriskyi and co-workers [Demirskyi D et al, 2012] on micrometre sized spherical balls of WC did reveal anomalous diffusion at the inter-particle neck regions. The group conducted some fundamental studies to understand the sintering mechanism in WC during SPS, conventional sintering and Microwave sintering. They measured both the linear shrinkage during the initial stage and also the neck growth rate (by SEM observations). Using the experimentally observed neck growth rate, they calculated the diffusion coefficient and arrived at the surprising result that the diffusion coefficient during SPS was *orders of magnitude higher* than during conventional sintering. This anomalous behaviour was attributed to a highly active surface and recrystallization at the neck region was clearly observed by SEM. Another recent report is from the group of Guyot and co-workers [Guyot P *et al*, 2012], who demonstrated micro welding at inter-particle areas of micron-sized Cu powders. They attribute this to the inductive coupling of the electromagnetic field leading to a decrease in electrical resistivity of the powders by several orders of magnitude (the *Branly effect*). These recent reports clearly suggest that particle overheating can lead to a GB complexion [Jian Luo, 2008, Shen Dillon *et al*, 2009] particularly during the initial stages of SPS.

Our observations of the low temperature sintered samples clearly show that the local field-induced temperature gradients can cause spontaneous melting and welding near the neck regions at temperatures as low as 1323 K. While the process of equilibrium melting (i.e., melting WC particles by slow heating to their melting point) may be expected to increase the activation energy for sintering (component of ΔH arising from the phase transformation of solid to liquid), this surface melting that is expected to occur in SPS is a totally non-equilibrium melting phenomenon which occurs rapidly. As Chaim [Rachman Chaim, 2007] and Chaim and Reinharz [Chaim R and Reinharz Bar-Hama, 2010] have suggested, this GB complexion leads to a plastically softened surface layer which can activate rapid atomic diffusion and promote particle rearrangement and creep leading to very low activation energies during sintering. At high temperatures, the 'particle' identity is lost and grain growth rate also increases. The thickness of the recrystallized layer decreases in comparison to the grain size, although the GB is still an active diffusion route until complete densification is achieved. This idea is consistent with our observed kinetic results and the microstructure.

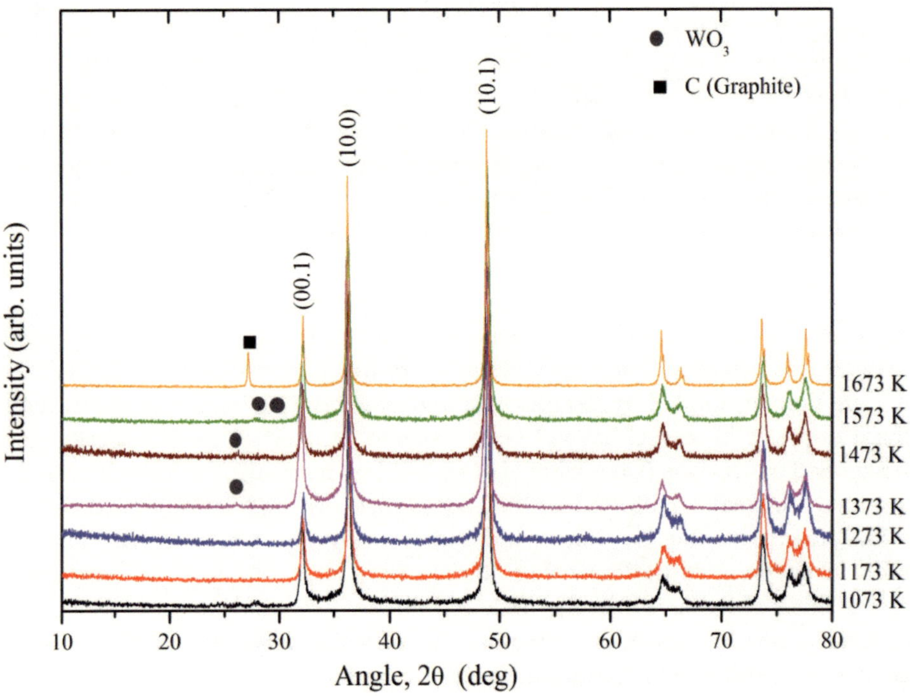

Figure 23. XRD of the samples interrupted at different temperatures. The WO_3 phase is seen at almost all low temperatures, while the final compact only shows WC and graphite. All the primary peaks are from the WC phase.

7. Conclusions

The SPS behaviour of *n*-WC appears to be a complex process involving size effects, field effects, chemical reactions and anomalously rapid diffusion. Experimental observations described in this work show evidence of planar defects, possible GB reconstruction and agglomeration that can contribute to the lowering of the sintering activation energy. In our analysis and from the current volume of literature cited to support our view, it is clear that the fundamental aspects of sintering pertaining to nano particles particularly in the presence of an electromagnetic field can be largely different, for which an exact theory is yet to be developed. A few significant results of this work are given below to summarize our findings:

1. The presence of excessive planar defects in the powder suggests that the quality of the nano powder is crucial for determining the sintering kinetics. In addition to defects, powder agglomeration controls sintering for most of the temperature range.
2. The low activation energies observed encourage efforts to consolidate nano powders to full density. However, not all temperatures are suitable for the sintering process, as agglomerates strongly impede densification at low temperatures. In those temperature ranges where agglomerates retard shrinkage, active surface diffusion and particle rearrangement acts to increase the compact density.
3. While at low temperatures the current assisted, over-heated surface is most likely the active diffusion route, at higher temperatures, grain growth acts to reduce the retarding effect of agglomerates leading to enhanced sintering.
4. The net sintering rate in the *n*-WC powder can be equated to the sum of three factors: $(\dot{\rho})_{total} = (\dot{\rho})_{defects} + (\dot{\rho})_{SD-PR} + (\dot{\rho})_{GB}$.

While the sintering mechanisms detailed in this work are not conclusive, it can be regarded as a pointer for furthering our understanding of the sintering behaviour of *n*-WC. The experimental observations do suggest that alternate, yet novel mechanisms are active during the SPS of *n*-WC, and certain factors that can be responsible have been discussed at length. However, a consistent theory of nano sintering specific to *n*-WC is still necessary. Such a description should therefore include the effects of GB plasticity and creeping induced by dislocation climb and glide in addition to the surface overheating phenomenon in nano materials during SPS. These points are motivated by the fact that even in a brittle material like WC, plasticity effects can be significantly enhanced as the 'GB phase' fraction increases.

Author details

A.K. Nanda Kumar
Dept. of Materials Science and Engineering, Case Western Reserve University,
Cleveland, Ohio,
USACentre for Advanced Research of Energy and Materials,
Faculty of Engineering, Hokkaido University, Sapporo, Japan

Kazuya Kurokawa
Centre for Advanced Research of Energy and Materials,
Faculty of Engineering, Hokkaido University, Sapporo, Japan

Acknowledgement

This work was carried out while AKNK was a foreign researcher at Hokkaido University, Japan. The project was partly funded by the Ohtaseiki Co., Ltd., Japan. AKNK also wishes to express his deep sense of gratitude to Prof. K Kurokawa, for having introduced him to this work and for providing financial support during his stay in Japan. Profs. A Yamauchi and N

Sakaguchi are also gratefully acknowledged for their timely help with the SPS and TEM work and also for many discussions during the course of this work.

8. References

[1] Agrawal D, Cheng J, Seegopaul P, Gao L, Grain growth control in microwave sintering of ultrafine WC-Co composite powder compacts, Powder Metall., 43 (2000) 15-16.

[2] Andon R J L, Martin J F, Mills K C and Jenkins T R, Heat capacity and entropy of tungsten carbide, I. Chem. Thermodynamics, 7 (1975) 1079-1084.

[3] Andrievky R A and Spivak I I, Strength of high melting point compounds - A Companion, Metallurgy, Moscow, (1983) [in Russian].

[4] Ashby M F, A first report on sintering diagrams, Acta Metall., 22 (1974) 275-289.

[5] Bacmann J J, Cizeron G, Dorn Method in the study of initial phase of uranium dioxide sintering, J. Am. Ceram. Soc., 51 (1968) 209–212.

[6] Bartha L, Atato P, Toth A L, Porat R, Berger S and Rosen A, Investigation of hip-sintering of nanocrystalline WC/Co powder, J. Adv. Mater., 32 (2000) 23-26.

[7] Beere W, The second stage sintering kinetics of powder compacts, Acta Metall., 23 (1975) 139-145.

[8] Bernard-Granger G and Guizard C, Spark plasma sintering of a commercially available granulated Zirconia powder I. Sintering path and hypotheses about the mechanism(s) controlling densification, Acta Mater., 55 (2007) 3493-3504.

[9] Breval E, Cheng J P, Agrawal D K , Gigl P, Dennis M and Roy R, Comparison between microwave and conventional sintering of WC/Co composites, Mat. Sci. Eng. A, 391 (2005) 285 -295.

[10] Buhsmer C P and Crayton P H, Carbon self-diffusion in tungsten carbide, J. Mater. Sci., 6 (1971) 981–988.

[11] Byung-Kwon Yoon, Bo-Ah Lee, Suk-Joong L. Kang, Growth behavior of rounded (Ti,W)C and faceted WC grains in a Co matrix during liquid phase sintering, Acta Mater., 53 (2005) 4677.

[12] Chaim R, Reinharz Bar-Hama O, Densification of nanocrystalline NiO ceramics by spark plasma sintering, Mat. Sci. Eng. A, 527 (2010) 462–468.

[13] Coble R L, Initial sintering of alumina and hematite, J. Am. Ceram. Soc., 41 (1958) 55–62.

[14] Coble R L, Sintering crystalline solids. I. Intermediate and final state diffusion models, J. Appl. Phys., 32 (1961a) 787–792.

[15] Coble R L, Sintering crystalline solids. II. Experimental test of diffusion models in powder compacts, J. Appl. Phys. 32 (1961b) 793–799.

[16] Cremer G D, (1944) *US Patent* No. 2,355,954.

[17] Demirskyi D, Hanna Borodianska, Dinesh Agrawal, Andrey Ragulya, Yoshio Sakka and Oleg Vasylkiv, Peculiarities of the neck growth process during initial stage of spark-plasma, microwave and conventional sintering of WC spheres, J. Alloy. Compd., 523 (2012) 1– 10.

[18] Di Yang, Rishi Raj and Hans Conrad, Enhanced sintering rate of zirconia (3Y-TZP) through the effect of a weak dc electric field on grain growth, J. Am. Ceram. Soc., 93 (2010) 2935–2937.

[19] Dominguez O and Bigot J, Material transport mechanisms and activation energy in nanometric Fe powders based on sintering experiments, Nanostruct. Mater., 6 (1995) 877-880.

[20] Duran P, Villegas M, Capel F, Recio P and Moure C, Low-temperature sintering and microstructural development of nanocrystalline Y-TZP powders. J. Eur. Ceram. Soc., 16 (1996) 945-952.

[21] Eugene A. Olevsky, Ludo Froyen, Impact of thermal diffusion on densification during SPS, J. Am. Ceram. Soc., 92 [S1] (2009) S122–S132.

[22] Giovanni Maizza, Salvatore Grasso, Yoshio Sakka, Tetsuji Noda and Osamu Ohashi, Relation between microstructure, properties and spark plasma sintering (SPS) parameters of pure ultrafine WC powder, Sci. Tech. Adv. Mater., 8 (2007) 644–654.

[23] Goren-Muginstein G R, Berger S and Rosen A, Sintering study of nanocrystalline WC powders, Nanostruct. Mater. 10 (1998) 795–804.

[24] Grebenkina V G and Denbnovetskaya E N, Temperature coefficient of electrical resistivity of some complex metal carbides, Translated from Poroshkovaya Metalturgiya (*in Russian*), 63 (1968) 34-36.

[25] Gupta V K , Dang-Hyok Yoon, Harry M. Meyer III and Jian Luo, Thin intergranular films and solid-state activated sintering in nickel doped tungsten, Acta Mater., 55 (2007) 3131–3142.

[26] Gurland J, Application of dihedral angle measurements to the microstructure of cemented carbides WC–Co, Metallography, 10 (1977) 461–468.

[27] Guyot P, Rat V, Coudert J F, Jay F, Maitre A and Pradeilles N, Does the Branly effect occur in spark plasma sintering?, J. Phys. D: Appl. Phys. 45 (2012) 092001-092005.

[28] Hulbert D M, Anders A, Andersson J, Lavernia E J and Mukherjee A K , A discussion on the absence of plasma in spark plasma sintering, Scripta Mater., 60 (2009) 835–838.

[29] Hulbert D M, Anders A, Dudina D V, Andersson J, Jiang D, Unuvar C, Anselmi-Tamburini U, Lavernia E J and Mukherjee A K, The absence of plasma in 'spark plasma sintering', J. Appl. Phys., 104 (2008) 033305.

[30] Itatani K, Itoh A, Howell F S, Kishioka A and Kinoshita M, Densification and microstructure development during the sintering of sub-rnicrometre magnesium oxide articles prepared by a vapour-phase oxidation process, J. Mater. Sci., 28 (1993) 719-728.

[31] Jian Luo, Liquid-like interface complexion: From activated sintering to grain boundary diagrams, Curr. Opin. Solid St. Mater. Sci., 12 (2008) 81–88.

[32] Jiang X X, Huang Dong-Shen and Weng Lu Quian, Sintering characteristics of microfine zirconia powder, J. Mater. Sci., 29 (1994) 121-124.

[33] Johnson D L and Cutler I B, Diffusion sintering: I, Initial stage sintering models and their application to shrinkage of powder compacts, J. Amer. Ceram. Soc., 46 (1963a) 541-545.

[34] Johnson D L and Cutler I B, Diffusion sintering: II, Initial sintering kinetics of Alumina, J. Amer. Ceram. Soc., 46 (1963b) 545-549.

[35] Kellett B J and Lange F F, Thermodynamics of densification. I. Sintering of simple particle arrays, equilibrium configurations, pore stability and shrinkage, J. Am. Ceram. Soc., 72 (1989) 725–741.

[36] Kim H C, Oh D Y and Shon I J, Sintering of nanophase WC-15 Vol.% Co hard metals by rapid sintering process, Int. J. Refract. Met. Hard Mater., 22 (2004) 197 – 203.

[37] Kinemuchi Y and Watari K, Dilatometer analysis of sintering behavior of nano-CeO_2 particles, J. Eur. Ceram. Soc., 28 (2008) 2019–2024.

[38] Kingery W D and Berg M, Study of initial stages of sintering solids by viscous flow, evaporation-condensation and self-diffusion, J. Appl. Phys., 26 (1955) 1205-1212.

[39] Kingery W D, Bowen H K and Uhlmann D R, Introduction to Ceramics, second ed., John Wiley and Sons, New York, 1975.

[40] Knorr P, Nam J G and Lee J S, Sintering behaviour of nano crystalline γ-NiFe powders, Metall. Mater. Trans. A, 31A (2000) 506-510.

[41] Kumar A K N, Watabe M, Yamauchi A, Kobayashi A and Kurokawa K, Spark plasma sintering of binderless n-WC and n-WC-X (X=Nb, Re, Ta, Ti, B, Si), Trans. JWRI, 39 (2010) 47-56.

[42] Lanfredi S, Dessemond L and Martins Rodrigues A C, Dense ceramics of $NaNbO_3$ produced from powders prepared by a new chemical route, J. Eur. Ceram. Soc., 20 (2000) 983-990.

[43] Lange F F, Sinterability of agglomerated powders, J. Am. Ceram. Soc., 67 (1984) 83–89.

[44] Li J G and Sun X, Synthesis and sintering behaviour of a nano crystalline α-alumina powder, Acta mater., 48 (2000) 3103-3112.

[45] Li T, Li Q, Lu L, Fuh J Y H and Yu P C, Abnormal grain growth of WC with small amount of cobalt, Philos. Mag. A, 87 (2007) 5657-5671.

[46] Luca Girardini, Mario Zadra, Francesco Casari and Alberto Molinari, SPS, binderless WC powders and the problem of subcarbide, Met. Powder Rep., (2008) 18-22.

[47] Munir Z A, Anselmi-Tamburini U and Ohyanagi M, The effect of electric field and pressure on the synthesis and consolidation of materials: A review of the spark plasma sintering method, J. Mater. Sci., 41 (2006) 763–777.

[48] Nabarro F R N, Bartolucci Luyckx S and Waghmare U V, Slip in tungsten monocarbide II. A first-principles study, Mater. Sci. Eng. A, 483-484 (2008) 9-12.

[49] Nobre M A L, Longo E, Leite E R and Varela J A, Synthesis and sintering of ultrafine NaNbO powder by use of polymeric precursors. Mater. Lett., 28 (1996) 215-220.

[50] Panda P C, Mobley W M and Raj R, Effect of the Heating Rate on the Relative Rates of Sintering and Crystallization in Glass, J. Am. Ceram. Soc., 72 (1989) 2361-2364.

[51] Rachman Chaim, Densification mechanisms in spark plasma sintering of nanocrystalline ceramics, Mater. Sci. Eng. A, 443 (2007) 25–32.

[52] Rahman M N, Ceramic Processing and Sintering, 2nd ed., Marcel Dekker, New York, (2003).

[53] Raj R and Bordia R K, Sintering behavior of bi-modal powder compacts, Acta Metall., 32 (1984) 1003-1019.

[54] Salvatore Grasso, Yoshio Sakka and Giovanni Maizza, Electric current activated/assisted sintering (*ECAS*): a review of patents 1906–2008, Sci. Technol. Adv. Mater., 10 (2009) 053001.

[55] Shen J. Dillon, Martin P. Harmer, and Jian Luo, Grain Boundary Complexions in Ceramics and Metals: An Overview, JOM, 61 (2009) 38-44.

[56] Shi, J. L., Lin, Z. X., Qian, W. J. and Yen, T. S., Characterization of agglomerate strength of coprecipitated superfine zirconia powders. J. Eur. Ceram. Soc., 13 (1994) 265-273.

[57] Swinkels F B and Ashby M F, A second report on sintering diagrams, Acta Metall., 29 (1980) 259-281.

[58] Theunissen G S A M, Winnubst A J A and Burggraaf A J, Sintering kinetics and microstructure of nanoscale Y-TZP ceramics, J. Eur. Ceram. Soc., 11 (1993) 315-324.

[59] Tokita M, Mechanism of Spark Plasma Sintering, in Proceedings of the International Symposium on Microwave, Plasma and Thermomechanical Processing of Advanced Materials, ed. S. Miyake and M. Samandi. JWRI, Osaka University, Japan (1997) 69–76.

[60] Tokita M, Suzuki S and Nakagawa K (2007) *Euro. Patent* No. EP1839782.

[61] Victor Zamoraa, Angel L. Ortiz, Fernando Guiberteau and Mats Nygren, Spark-plasma sintering of ZrB2 ultra-high-temperature ceramics at lower temperature via nanoscale crystal refinement, J. Eur. Ceram. Soc., 32 (2012) 2529–2536.

[62] Wang J and Raj R, Estimate of the activation energies for boundary diffusion from rate controlled sintering of pure alumina and alumina doped with zirconia and titania, J. Am. Ceram. Soc., 73 (1990) 1172-1175.

[63] Wenming Zeng, Lian Gao, Linhua Gui and Jinkun Guo, Sintering kinetics of α-Al_2O_3 powder, Ceram. Int., 25 (1999) 723-726.

[64] Yann Aman, Vincent Garnier and Elisabeth Djurado, Spark plasma sintering kinetics of pure α-Alumina, J. Am. Ceram. Soc., 94 (2011) 2825–2833.

[65] Young W S and Cutler I B, Initial sintering with constant rates of heating, J. Am. Ceram. Soc., 53 (1970) 659–663.

Chapter 3

Tungsten Carbide as an Addition to High Speed Steel Based Composites

Marcin Madej

Additional information is available at the end of the chapter

http://dx.doi.org/10.5772/51243

1. Introduction

Design criteria for high strength tool materials have to include wear resistance of the abrasive particle, high hardness and adequate toughness. Cold compaction and vacuum sintering of PM high speed steels (HSSs) to full density is now a well established technique [1-3]. In recent years, work has been undertaken to sinter metal matrix composites that contain ceramic particles in HSSs by the same route. Most studies have focused on sintering with additions of hard ceramics such as Al_20_3, VC, NbC, TiC, WC and TiN with the aim of producing a more wear resistant HSS type material [4-18]. These composite materials have been developed for wear resistance applications as attractive alternative to the more expensive cemented carbides. Compared with high strength steels, these composite materials have higher hardness, wear resistance and elastic modulus. However, depending on size and distribution, the addition of brittle ceramic particles may cause degradation of bend strength and toughness owing to the initiation of cracks at or near the reinforcing particles. In order to ensure good bonding at the ceramic/matrix interfaces, the ceramic particles must exhibit some reactivity with the matrix. In contrast to Al_2O_3, which presents no interface reactions with the iron matrix, the diffusion of iron from the matrix into the MC carbide particles establishes a good cohesion across the ceramic/matrix interface. Besides, these carbides are stable in contact with iron during sintering and do not dissolve extensively. Therefore, MC particles were chosen as the reinforcement. A cheap and easy route to develop high speed steels reinforced with MC carbides consists of mixing powders of commercial high speed steel powders with the carbides.

There are two methods by which tungsten carbide powders are produced from the tungsten-bearing ores. Traditionally, tungsten ore is chemically processed to ammonium paratungstate and tungsten oxides. These compounds are then hydrogen-reduced to tungsten metal powder. The fine tungsten powders are blended with carbon and heated in a hydrogen atmosphere between 1400 and 1500 °C (2500 and 2700 °F) to produce tungsten carbide particles with sizes varying from 0.5 to 30 μm. Each particle is composed of numerous tungsten carbide crystals. Small amounts of vanadium, chromium, or tantalum are sometimes added to tungsten and carbon powders before carburization to produce very fine (<1 μm) WC powders. In a more recently developed and patented process, tungsten carbide is produced in the form of single crystals through the direct reduction of tungsten ore (sheelite). The melting point of hexagonal tungsten carbide WC is equal to ~2780°C which is less than melting point of nonstoichiometric TiC, ZrC, HfC, NbC, and TaC carbides [1,2]. Hardness of tungsten carbide WC is equal to 18-22 GPa at room temperature which is less than that of nonstoichiometric TiC_y, ZrC_y, HfC_y, VC_y, NbC_y and TaC_y carbides [1, 2]. However, the hardness of WC is sufficiently stable in a wide temperature interval. For example, the microhardness *HV* of hexagonal WC carbide decreased from ~18 to ~12GPa at heating from room temperature to 1000°C whereas the HV of nonstoichiometric transition metal carbides decreased under the same conditions from their maximum values to 3-8GPa. A thermal expansion coefficient of WC is equal to $\sim 5.5 \cdot 10^{-6}$ K^{-1} [4-10] and is half as much as that of other transition metal carbides.

Metal matrix composites MMC have appeared as a bright option for wear applications. These materials combine a soft metallic matrix with hard ceramic particles that withstand wear. Different matrixes were studied, and aluminum, stainless steel and high speed steel HSS. Matrix composites showing outstanding wear behaviour. High hardness, mechanical strength, heat resistance and wear resistance of high speed steel (HSS) make it an attractive material for manufacture MMC. High speed steels comprise a family of alloys mainly used for cutting tools. Their name – high speed steel – is a synthesis of the following two features [11]:

a. the alloys belong to the Fe–C–X multicomponent system, where X represents a group of alloying elements in which Cr, W or Mo, V, and Co are the principal ones;
b. the alloys are characterized by their capacity to retain a high level of hardness even when submitted to elevated temperatures resulting from cutting metals at high speed.

The carbides are predominantly formed from the strong carbide formers V, W, Mo and Cr. Depending on the composition and on the thermal parameters (sintering temperature and cooling rate) several types of carbides can be formed. The main types are M_6C, M_2C, MC and $M_{23}C_6$. Their formation is a result of cooling rate and alloy composition. The carbides can be formed directly from the melt during solidification, by eutectic reaction or by decomposition of other types of carbide. The latter case can be seen from the decomposition of M_2C carbide into M_6C and MC carbides when HSS is annealed at temperatures above 1000°C. The importance of the alloying elements in HSS is the effect they have on the type of carbide formed and the temperature of formation, and also the effect they have on the tempering of martensite. The types of carbides are different in crystal structure and composition. The carbides MC, M_6C and $M_{23}C_6$ have a fcc crystal structure, while the metastable M_2C has a hcp structure. The carbides are also different in composition with higher V and Ti levels and lower Fe and Cr levels in MC compared with M_6C and M_2C, which are rich in Mo and W. The carbide $M_{23}C_6$ is Cr rich, with high solubility for Fe and low solubility for W and Mo. Iron can substitute Cr when W and Mo are dissolved in this carbide. Annealed HSS generally contains $M_{23}C_6$. Considering HSS grades such as M3/2 it is clear that primary M_6C formation and growth along cell boundaries is a key problem. If the M_6C formation can be limited, while sintering to near full density is maintained, an acceptable final microstructure can result. The alloy composition of HSS may favour the formation of a certain type of carbide. For a given amount of V, increasing the amount of W favours the formation of M_6C at the expense of MC. A high amount of Mo has the opposite effect. The carbides that form by eutectic reaction increase in volume fraction with increasing C content, higher W/Mo ratio in the case of eutectic M6C, the V content9, and decreasing cooling rate. Many studies demonstrate how the as sintered microstructure of HSS can be tailored by means of addition of elemental powder to gas atomised M3/2 powder. The experiments include the addition of elemental Si to repel C from the melt when sintering, the addition of elemental Ni to avoid the formation of pearlite and stabilize austenite, and finally the addition of elemental V to form MC on the expense of cementite. Additions of carbides to high speed steels have been studied by a number of authors [9-19]. Thermodynamically less stable carbides (SiC, Cr_3C_2) easily dissolve in the high speed steel matrix during sintering or annealing and are not retained as discrete hard particles. Intermediate carbides such as WC, VC, Mo_2C and NbC, which include elements that are alloyed to high speed steel react with the steel matrix to produce new carbide phases with compositions similar to those of the normal primary carbides present in high speed steel, e.g M_6C {Fe_3W_3C or Fe_eMo_3C} and niobium or vanadium rich MC type carbides. Thermodynamically stable carbides such as TiC are retained more less in their original form but also encourage MC carbides to form within the steel matrix and the steel/TiC interface because of some slight dissolution of the TiC particles. The TiC additions decreases sinterability by raising the sintering temperature required to achieve full density in a number of grades high speed steels, including M3/2. Most authors report also that the TiC addition reduces bend strength of the HSS and causes some slight increase in hardness.

The use of powder metallurgy PM techniques for manufacturing HSS is on the increase. In addition to the typical advantages of PM raw material savings, low energy costs. PM HSS present better microstructural features than conventional wrought steels homogeneity of carbide distribution in the matrix and smaller grain and carbide sizes, among others. These advantages mean an improvement of properties. PM techniques allow a higher content of alloying elements and the addition of the ceramic particles previously mentioned. If conventional compacting or CIP is used as the powder forming technique, a sintering process is then necessary. Sintering is usually carried out under vacuum conditions at temperatures below solidus temperature, although nitrogen-based atmospheres can sometimes be used, but this process does not completely eliminate porosity. A technique that provides densities as high as HIP but at a lower cost is liquid phase sintering. Liquid phase can be obtained in three ways:

1. by addition of a compound with a lower melting point than the material to be sintered, or that forms a eutectic,
2. by supersolidus sintering, which consists of heating to temperatures above the solidus temperature of the material to be sintered,
3. by infiltration technique [20, 21].

Infiltration is basically defined as "a process of filling the pores of a sintered or unsintered compact with a metal or alloy of a lower melting point." In the particular case of copper infiltrated iron and steel compacts, the base steel matrix, or skeleton, is heated in contact with the copper alloy to a temperature above the melting point of the copper, normally within the range of 1095° to 1150°C [22, 23]. Through capillary action, the molten copper alloy is drawn into the interconnected pores of the skeleton and ideally fills the entire pore volume. Filling of the pores with higher density copper can result in final densities in excess of 95% of the composite theoretical value. Completely filled skeletons also allow for secondary operation such as pickling and plating without damaging the structure through internal corrosion. Pressure tight infiltrated components are also possible for specific applications that demand the absence of interconnected porosity. The infiltration process is generally subdivided into two fundamental methods: single step or double step. The single step or single pass is presently the preferred infiltration method that consists of one run or passes through the furnace. In this process, the unsintered (green) steel and copper alloy compacts are placed in contact prior to furnace entry. The typical arrangement is to place the copper alloy infiltrant compact on the top surface of the steel compact. In some cases, it is preferred to place the steel compact on top of the infiltrant compact, or, infiltrate from top and bottom simultaneously. During the full furnace cycle, the steel base compact is ideally partially sintered prior to attaining the melting points of the infiltrant composition. Preferably, multi-independent zone furnaces are employed allowing for preheat, or lubricant burn-off, followed by pre-sintering (graphite solution) and finally infiltration. The double step or double pass infiltration method consists of pre-sintering or full sintering of only the steel compact in one pass through the furnace. After the first sintering pass, the unsintered (green) infiltrant compact is placed in contact with the sintered steel part, and the full furnace cycle is repeated. The infiltrating powders available may be used for both the single and double step processes. Most, if not all, infiltrating powders are prepared as a pre-blended and/or a pre-lubricated lot or batch and are designed for typical compacting operations. Shapes of infiltrant compact forms vary substantially depending upon the amount required and the configuration of the steel skeleton. Usually, simple infiltrant shapes, such as bars, cylindrical slugs, or annuli are compacted to a specific weight and are placed on the iron components in single or multiple contact arrangements.

The M3/2 high speed steel reinforced with tungsten carbide and infiltrated with copper was chosen for this investigation. The present paper describes and discusses the microstructural characteristics and mechanical behaviour of the composite system M3/2-WC-Cu.

2. Experimental procedure

2.1. The powders

Water atomised M3 grade 2 powder of - 160μm were obtained from POWDREX SA in the annealed condition. This powder was chosen, in preference to the more commonly used M2, for its better compressibility and sinterability. Chemical composition of this powder is given in the Table 1. As reinforcement's commercial tungsten carbides of - 3μm were used. The powder properties, including electrolytic copper are given in Table 2.

compound	C	Cr	Co	Mn	Mo	Ni	Si	V	W	Fe	O
composition	1,23	4,27	0,39	0,21	5,12	0,32	0,18	3,1	6,22	rest	626ppm

Table 1. Chemical composition of M3/2 HSS powders, wt-%

Powder properties	M3/2	WC	Cu
Bulk density, g/cm^3	2,26	2,70	1,60
Flow time, s/50g	38,5	-	-
Densification at 600MPa, g/cm^3	6,08	6,71	6,90
Particle size, μm	0÷160	0÷3	0÷40

Table 2. Properties of the used powders

The powders morphology is shown in fig 1.

Figure 1. Scanning electron microscopy (SEM) morphology of powder particles:
a) high speed steel M3/2 class, b) tungsten carbides WC

It can be seen that the microstructure of the M3/2 grade HSS powders consists of a thin carbides in a martensitic-bainitic matrix. MC carbides being the white ones, Chile M C carbides are grey. Typical microhardness values for a powder is $\mu HV_{0,065} = 284 \pm 17$.

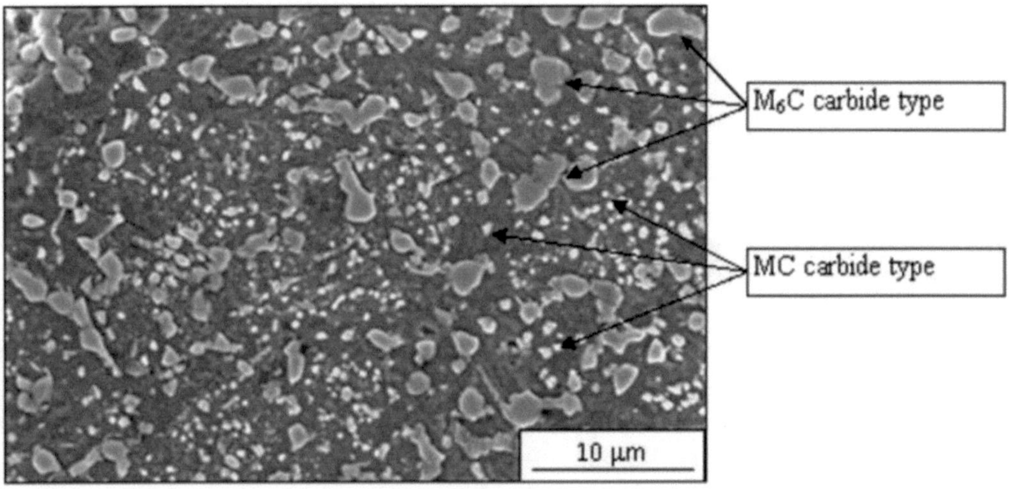

Figure 2. The microstructure of M3/2 grade HSS powder, SEM

2.2. Experimental technique

The compositions of powder mixtures are:

1. 100% M3/2,
2. M3/2 + 10% WC,
3. M3/2 + 30%WC.

Composite mixtures were blended in a Turbula® T2F blender for 30 min. The M3/2 powder and composite powders were uniaxially cold compacted in a cylindrical die at 800 MPa.

The infiltration process is subdivided into two fundamental methods: single step or double step. In the single step the unsintered (green) high speed steels or composite mixtures and copper alloy compacts were placed in contact prior to vacuum furnace entry. The copper alloy infiltrant compacts were placed on the top surface of the green compacts. The double step infiltration method consists of pre-sintering of only the green compacts. After the first sintering process, the infiltrant compacts (specify weight copper green compacts) were placed on the top surface of the sintered composites and were placed to vacuum furnace entry.

Half of green compacts were sintered in vacuum at 1150°C for 60 minutes. Sintered specimens and green compacts were analysed before infiltration. Density measurements via the Archimedes method were used to define the level of porosity.

Thereby obtained skeletons were subsequently infiltrated with copper, by gravity method, in vacuum furnace at 1150°C for 15 minutes. The infiltrated composites were cooled as fast as vacuum furnace.

The sintering and infiltration process was carried out in vacuum better than 10^{-2} Pa.

Densities of sintered materials were evaluated by a method based on the Archimedes principle, according to MPIF standard 42. Measured values were compared with theoretical values to obtain relative densities (ρ_t). The theoretical densities for the composites were calculated according to the expression (1):

$$r_t = (r_a \times X_a + r_b \times X_b) \quad (1)$$

where ρ_a is carbide density, X_a the volumetric fraction of WC carbide, ρ_b the density of high speed steel M3/2, and X_b the volumetric fraction of the HSS.

The MMC composites materials were characterized using various techniques. The infiltrated specimens were subsequently tested for mechanical properties. Therefore, mechanical properties were characterized by Brinell hardness, wear resistance test and three points bend tests in order to determine the influence of tungsten carbide. The microstructure of the composites was examined by means of both light microscopy (LM) and scanning electron microscopy (SEM). Characterization of microstructures and the identification of phases present were performed by both optical and scanning electron microscopy, assisted by the use of X-ray energy dispersive analysis (EDX), backscattered electron image contrast, and some X-ray diffraction data. Reaction temperatures were determined by dilatometric study.

The wear tests were carried out using the block-on-ring tester (Figure 3).

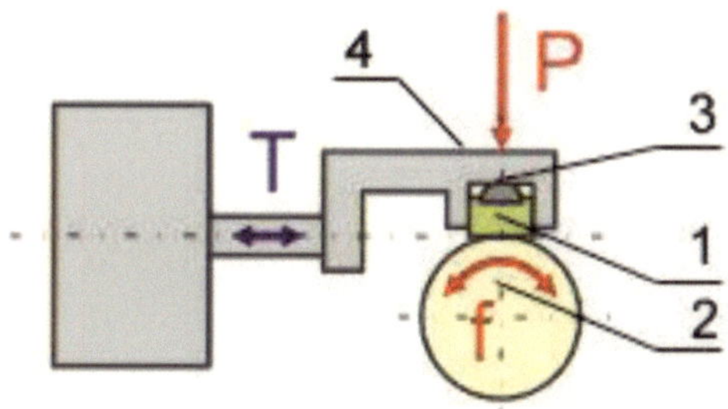

Figure 3. Tribosystem T05 - wear test principle

During the test a rectangular wear sample (1) was mounted in a sample holder (4) equipped with a hemispherical insert (3) ensuring proper contact between the test sample and a steel ring (2) rotating at a constant speed. The wear surface of the sample was perpendicular to the loading direction. Double lever system was used to force the sample towards the ring with the load accuracy of ±1%.

The wear test conditions were:

- test sample dimensions: 20 x 4 x 4 mm,
- rotating ring: heat treated steel 100Cr6, 55 HRC, ø49,5 x 8 mm,
- rotational speed: 500 rpm,
- load: 165 N,
- sliding distance: 1000 m.

The measured parameters were:

- loss of sample mass,
- friction force *F* (used to calculate the coefficient of friction).

The friction coefficient was measured continuously during the test, and the wear coefficient was calculated by means of the following expression (2):

$$F = \frac{friction \cdot force[N]}{load[N]} \tag{2}$$

Wear tracks were analyzed by LM to clarify wear mechanisms.

3. Results and discussion

3.1. Characterization of porous skeletons

The combined effects of tungsten carbide content and powder processing route on the relative density and shrinkage of the porous skeleton are shown in Figure 4 and 5.

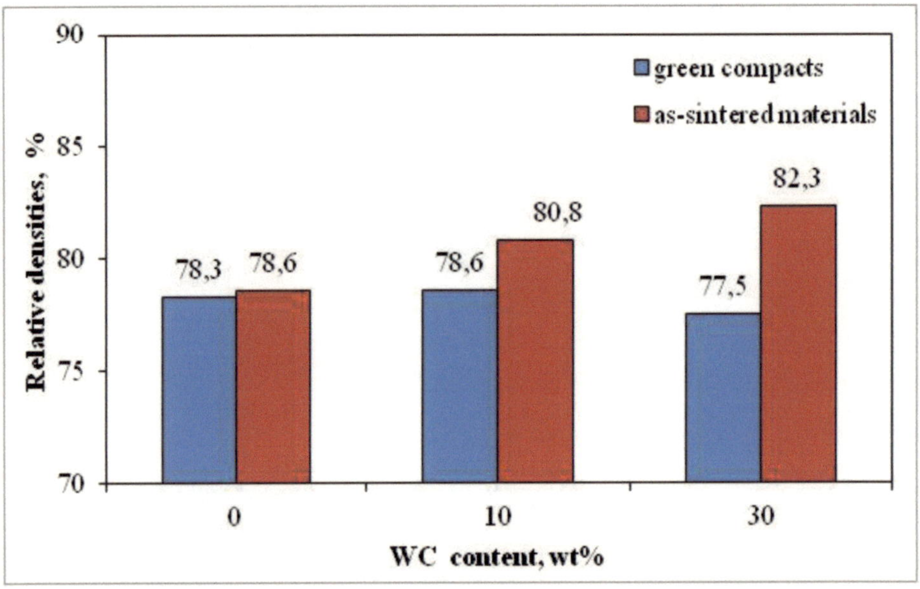

Figure 4. Relative densities of green compacts and pre-sintered porous skeletons as a function of tungsten carbide WC content

Figure 5 shows the effect of WC content on compressibility and shrinkage of high speed steel powders. It is evident that green density of compact decreases with increasing WC content. This attributes to hard and non-deforming nature of the tungsten carbide WC reinforcements, which constricts HSS-particle deformation, sliding and rearrangement during compaction. Additions of 30% tungsten carbide increase the as-sintered density.

Figure 4 shows that the M3/2 grade HSS cannot be fully densified at 1150°C, and that the as-sintered density is approximately equal to the green density. Additions of 30% tungsten carbide increase the as-sintered density presumably due to the occurrence of a liquid phase resulting from a chemical reaction occurring between the HSS matrix and tungsten carbide particles. As exemplified in Fig. 6, marked specimen expansion followed by its rapid contraction has indicated that the chemical reaction takes place at temperatures between 1080 and 1110°C.

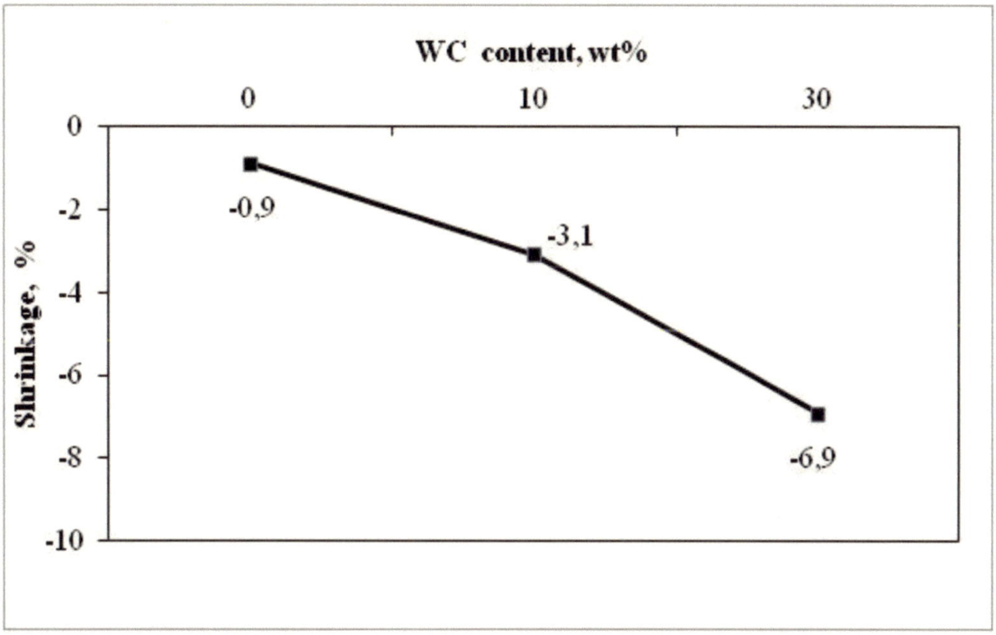

Figure 5. Shrinkage of compacts during sintering as a function of tungsten carbide WC content

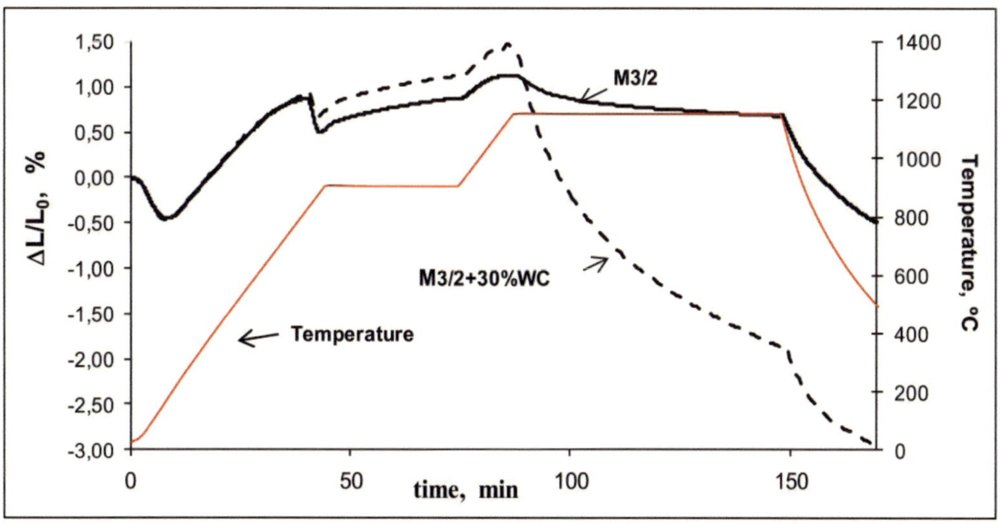

Figure 6. Dilatometric curves recorded on heating the M3/2 and HSS M3/2 + 30% WC material to the sintering temperature

Figures 4 and 5 show the morphologies of capillaries in both green compacts and pre-sintered skeletons.

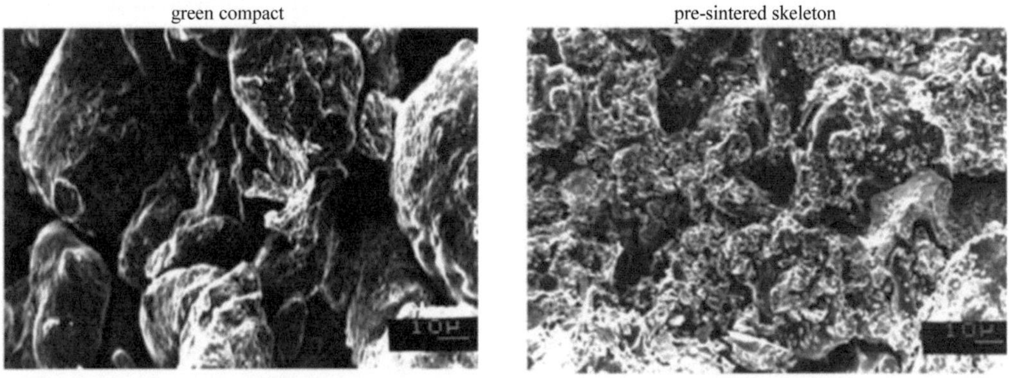

Figure 7. The morphologies of capillaries in M3/2 grade HSS, SEM

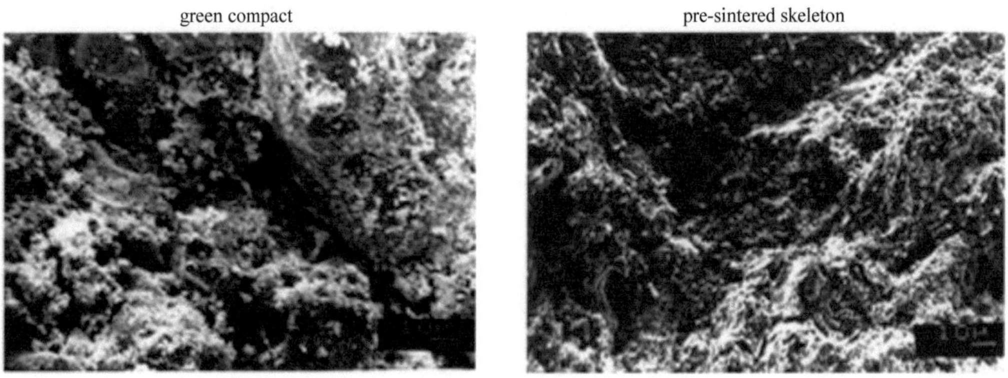

Figure 8. The morphologies of capillaries in M3/2 HSS + 30% WC , SEM

From the microstructural observations (Figures 7 and 5) it may be concluded that the morphologies of capillaries are mainly affected by the manufacturing route and powder characteristics (Fig. 1), such as powder particle size and morphologies of powder particles.

Finally, the porous skeletons were vacuum-infiltrated by gravity method at temperature: 1150°C for 15 min.

3.2. Characterization of as infiltrated materials

The as-infiltrated properties of the investigated composites are a complex function of the manufacturing route and tungsten carbide content. The properties of the as-infiltrated composites are shown in Figures 9÷12.

From Figure 9 it is evident that the molten copper is drawn into the interconnected pores of the skeleton, through a capillary action, and fills virtually the entire pore volume to yield final densities exceeding 97% of the theoretical value. In all cases, the additions of tungsten carbides aren't causes in the final density of the materials as compared with the base material.

Direct infiltration of as-sintered skeletons with copper results in the highest densities. This may be result of deoxidation powder particle surfaces during sintering in vacuum. Figure 10 show the final Cu content in as-infiltrated high speed steel based composites.

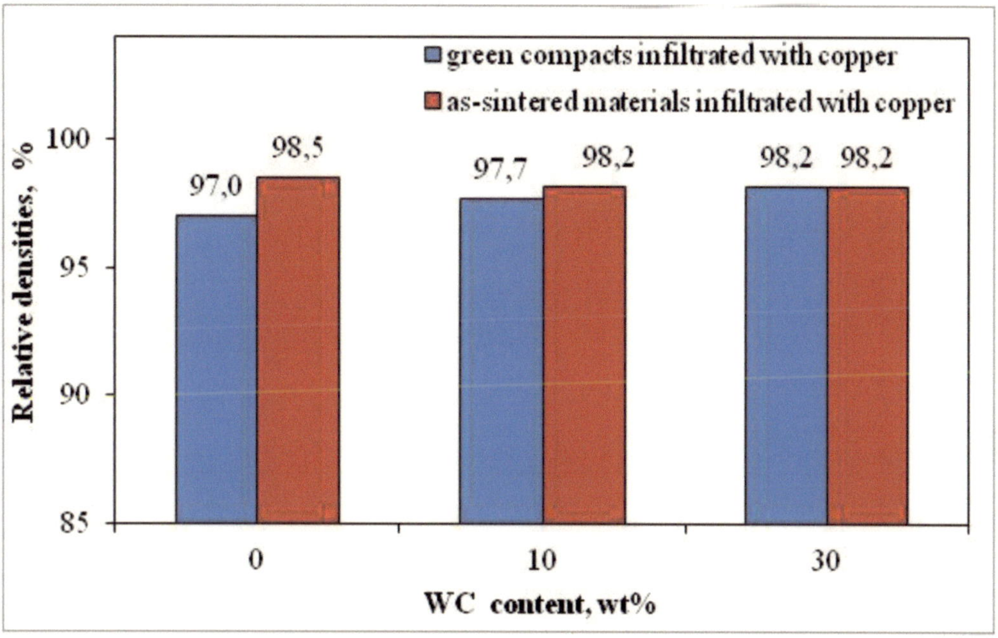

Figure 9. Relative densities of as-infiltrated composites

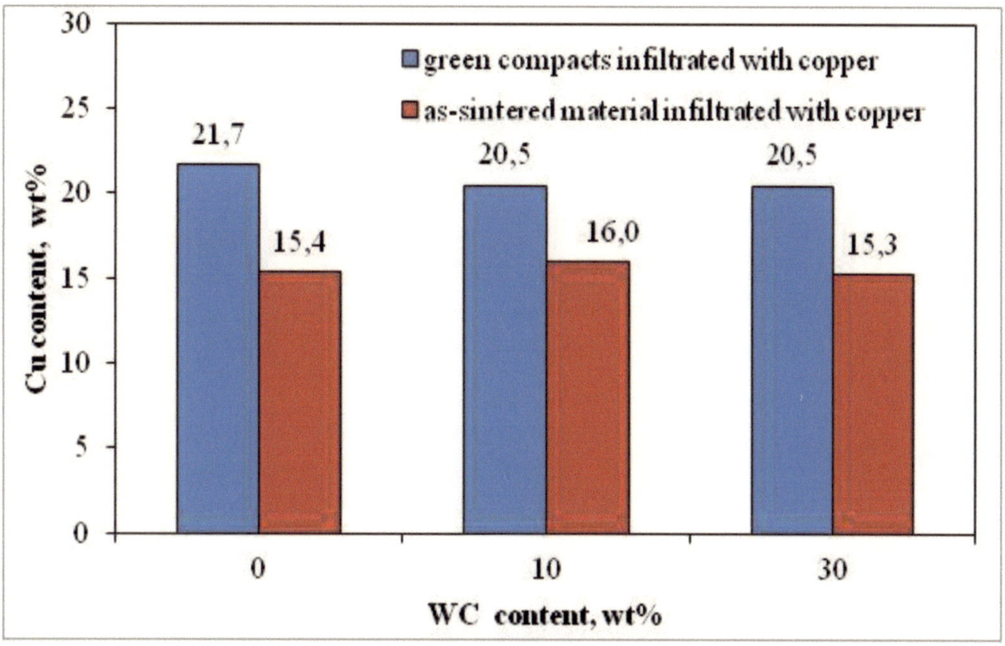

Figure 10. Copper content in as-infiltrated composites.

From Figure 10 it is evident that the copper content in as-sintered materials is lower than in infiltrated green compacts. It is affected by densification during sintering of porous skeletons.

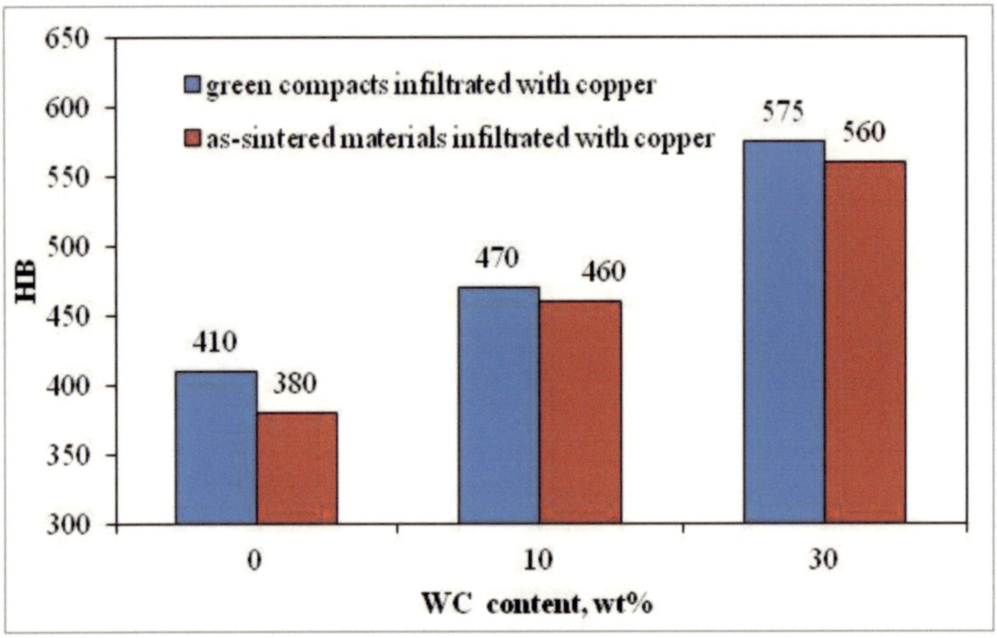

Figure 11. The Brinell Hardness of as-infiltrated composites

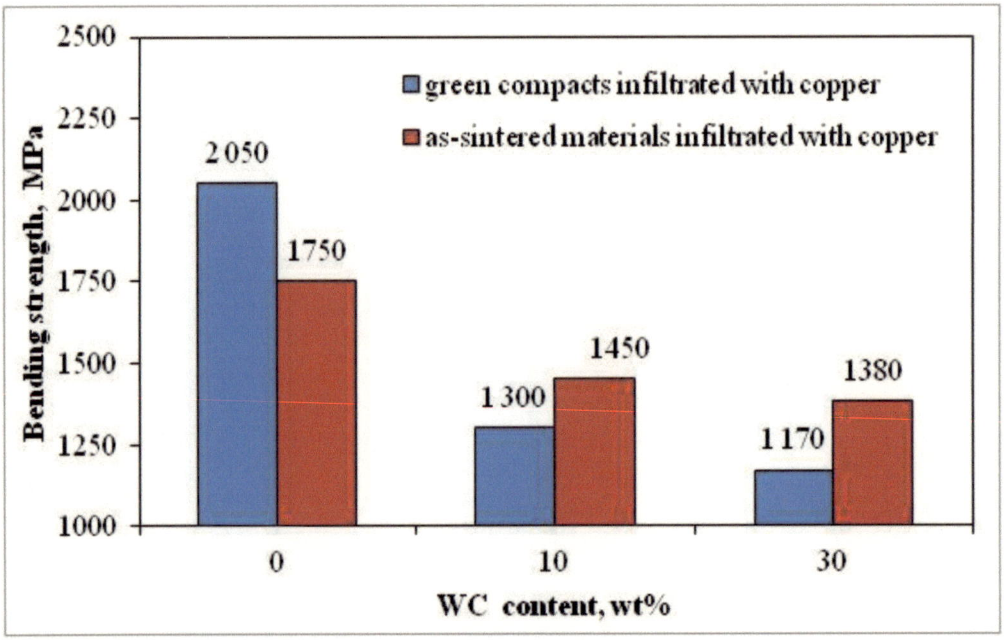

Figure 12. The Bending Strength of as-infiltrated composites

The Brinell hardness of the as-infiltrated composites increases with the percentage of additions of tungsten carbide WC. Considerable differences in hardness between the materials obtained from the two infiltration routes have been observed, with higher hardness numbers achieved with direct infiltration of green compacts, this can be explained by the diffusion of carbon and alloying of iron particles during sintering.

The Bending Strength of the as-infiltrated composites decreases with the increased content of tungsten carbide WC in the starting powder mix. For pre-sintered samples made of M3/2 – tungsten carbide mixture an increase of bending strength occurs; this can be explained by the chemical reaction between the tungsten monocarbide and HSS matrix.

3.3. Microstructures

Typical microstructures of a copper infiltrated green compact and pre-sintered skeleton are shown in Figures 13 – 15.

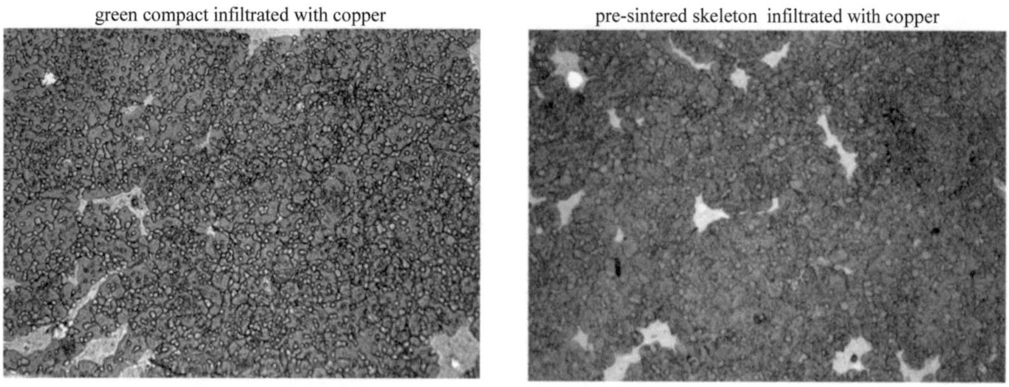

Figure 13. Microstructures of M3/2 HSS based composites

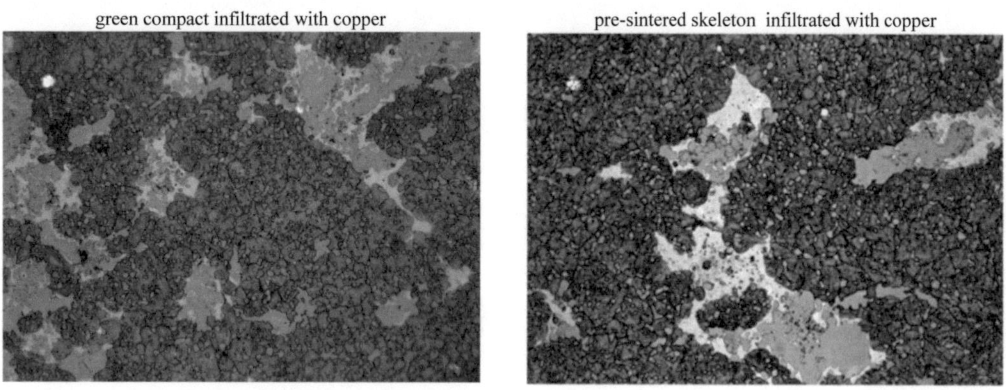

Figure 14. Microstructures of M3/2 HSS + 10%WC composites

It can be seen that the microstructure of the M3/2 grade HSS based composites consists of a steel matrix with finely dispersed carbides and islands of copper. Figures 13 and 15 shows tungsten carbides located within the grains and on the grain boundaries as well. Microstructures show small porosity at both sintering temperatures, and carbides are larger at the pre-sintered skeleton infiltrated with copper, MC carbides being the white ones, while MC carbides are grey. Some free copper areas are also present. In some places, these added carbides are related with white MC carbides, but free copper dark grey is preferentially linked to WC.

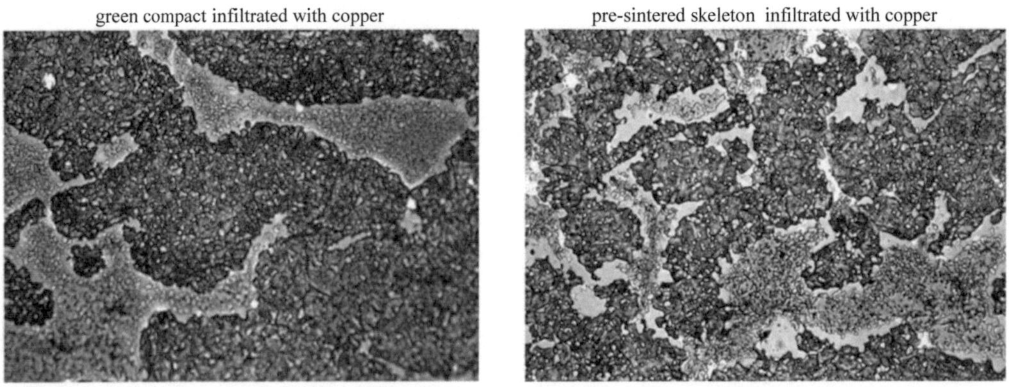

Figure 15. Microstructures of M3/2 HSS + 30%WC composites

SEM microstructures of a copper infiltrated green compact and pre-sintered skeleton are shown in Figures 16 – 17.

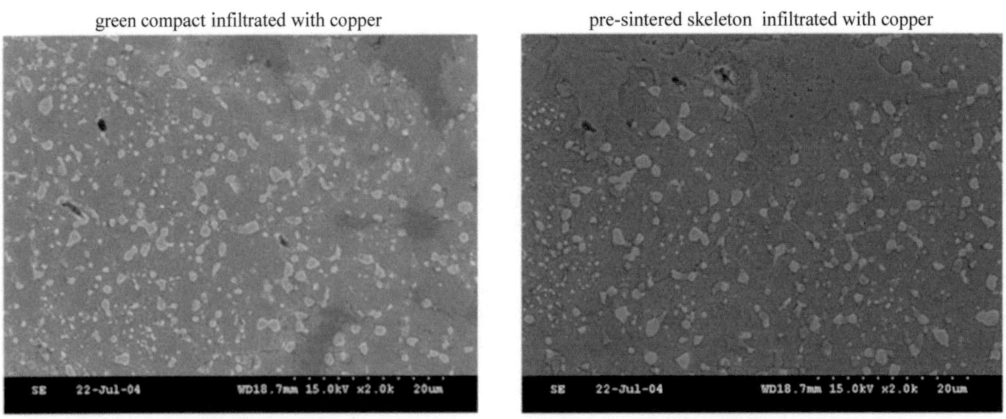

Figure 16. SEM microstructures of M3/2 HSS based composites

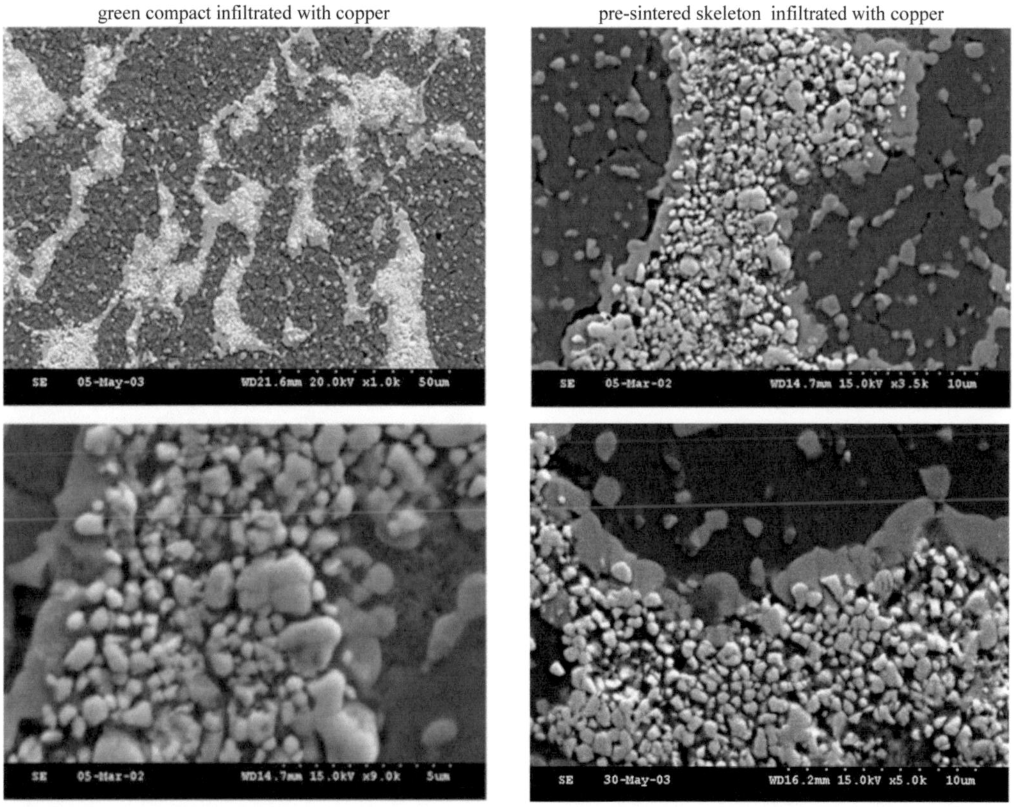

Figure 17. SEM microstructures of M3/2 HSS + 30%WC composites

4. Phase identification

Phase identification of the composites was performed by a Tur 62 X-ray diffraction (XRD) machine with Cu target (K_{α}, $\lambda = 1.5406Å$).

Figure 19 show the XRD patterns of samples M3/2 + 30%WC. They illustrate the existence of the main carbides M_6C and MC as well as the existence of ferrite and austenite and the high intensity for the main Cu peak in sample green compact infiltrated with copper compared with sample pre-sintered skeleton infiltrated with copper as well as the higher intensity coming from higher volume of copper in as-infiltrated green compact. It should be noted that the intensity of the Fe_3W_3C peaks in sample pre-sintered skeleton infiltrated with copper can be explained by the chemical reaction between the tungsten monocarbide and HSS matrix.

The SEM and EDX analysis performed on the specimens containing M3/2 10 and 30% tungsten carbide have revealed the carbide phase evenly distributed within the copper-rich regions. As it is apparent from Figures 20 and 21, WC reacts with the surrounding HSS matrix and forms a tungsten and iron-rich M_6C carbide grain boundary network during sintering of porous skeletons.

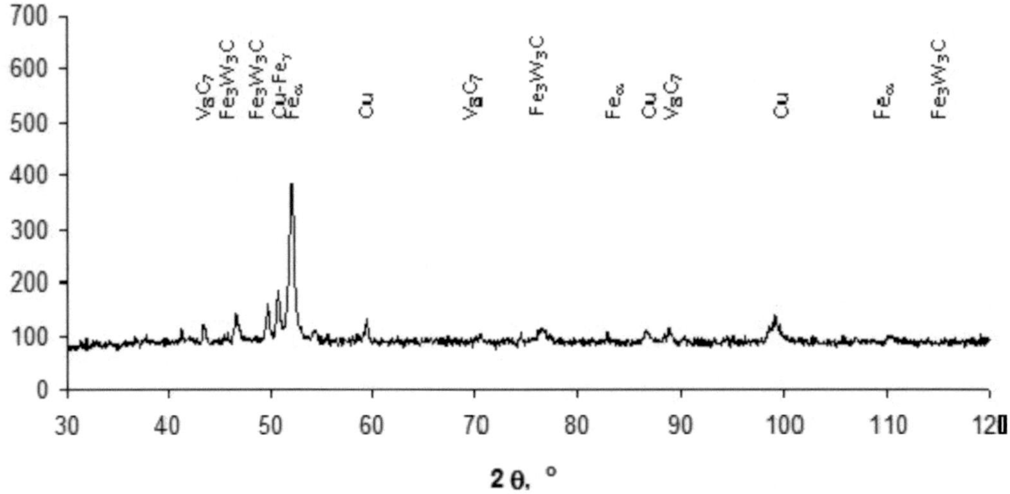

Figure 18. XRD pattern from M3/2 composites

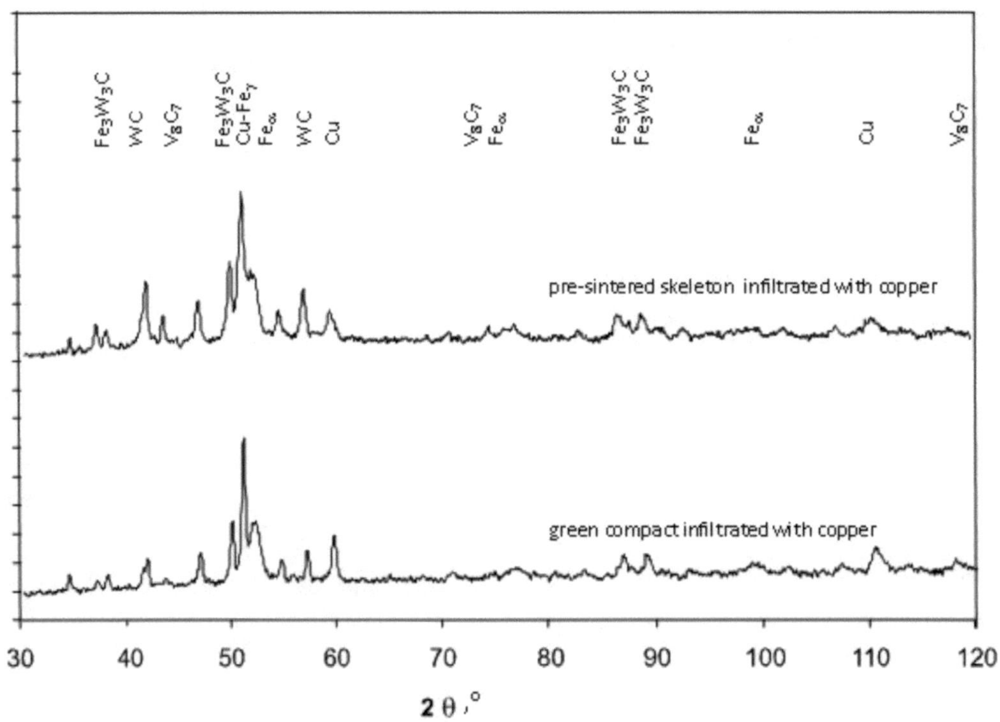

Figure 19. XRD pattern from M3/2 +30% WC composites

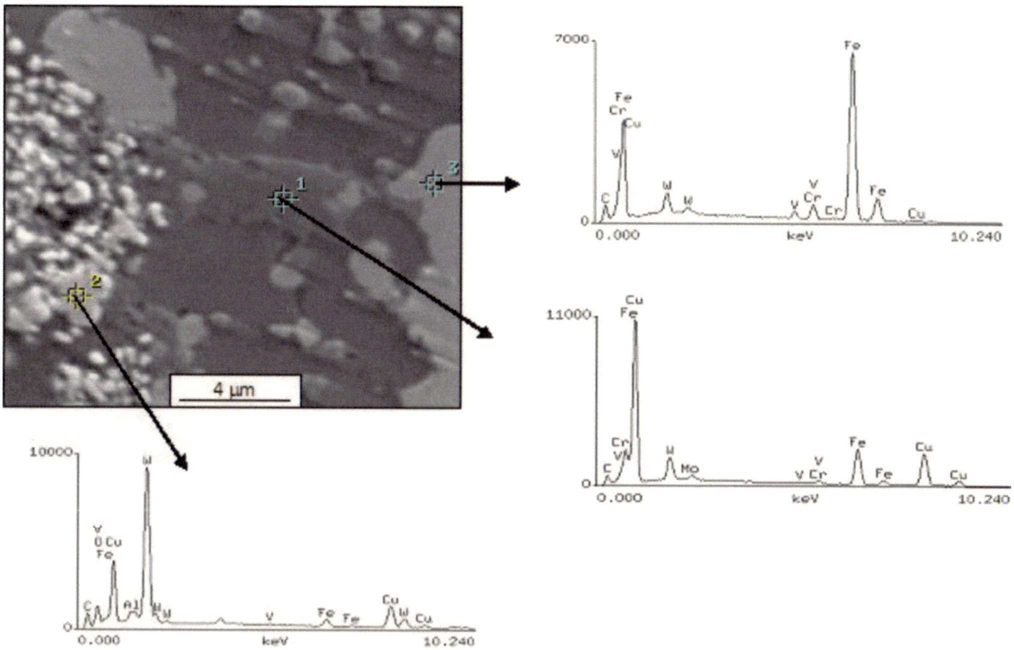

Figure 20. The microstructure of pre-sintered skeleton M3/2+30%WC infiltrated of copper and the qualitative EDX analysis, 1 – steel matrix, 2 – tungsten carbide WC, 3 – carbide M_6C type

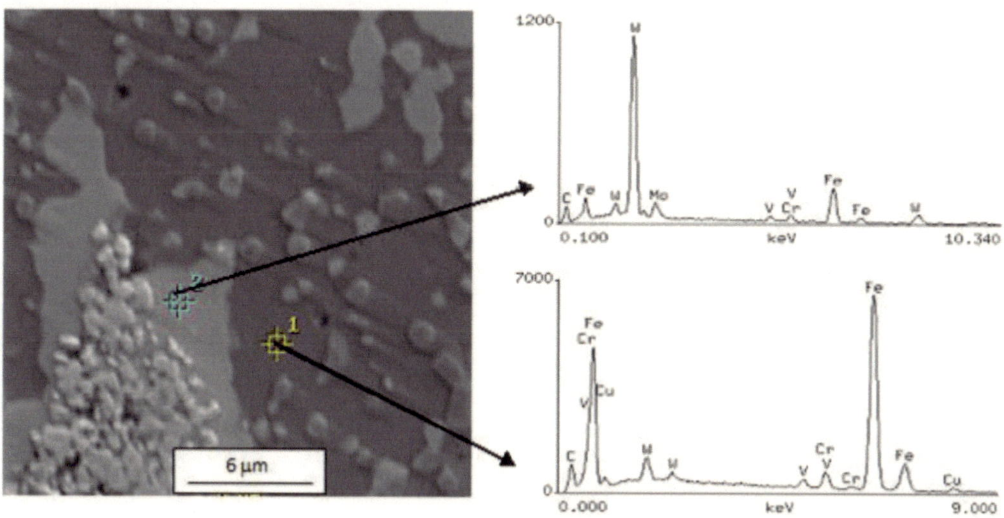

Figure 21. The microstructure of pre-sintered skeleton M3/2+30%WC infiltrated of copper and the qualitative EDX analysis, 1 – steel matrix, 2 – carbide M_6C type

Intermediate carbides such as WC which include elements that are alloyed to high speed steel react with the steel matrix to produce new carbide phases with compositions similar to those of the normal primary carbides present in high speed steel, e.g M_6C {Fe_3W_3C }. The EDS analysis was carried out on sample pre-sintered skeleton M3/2+30%WC infiltrated of copper to illustrate the chemistry of this carbides. Figure 22 shows elemental intensity maps for the alloying elements Fe, W and Cu. Most Fe is found in the matrix and grey M_6C carbides, while W is found in the whitish M_6C carbides. From Figure 21 it also evident that copper diffuse to steel matrix.

The carbide agglomerations observed are due to the non-assisted system used for mixing the powders; they could be avoided by using a more eficient mixing system such as a ball mill.

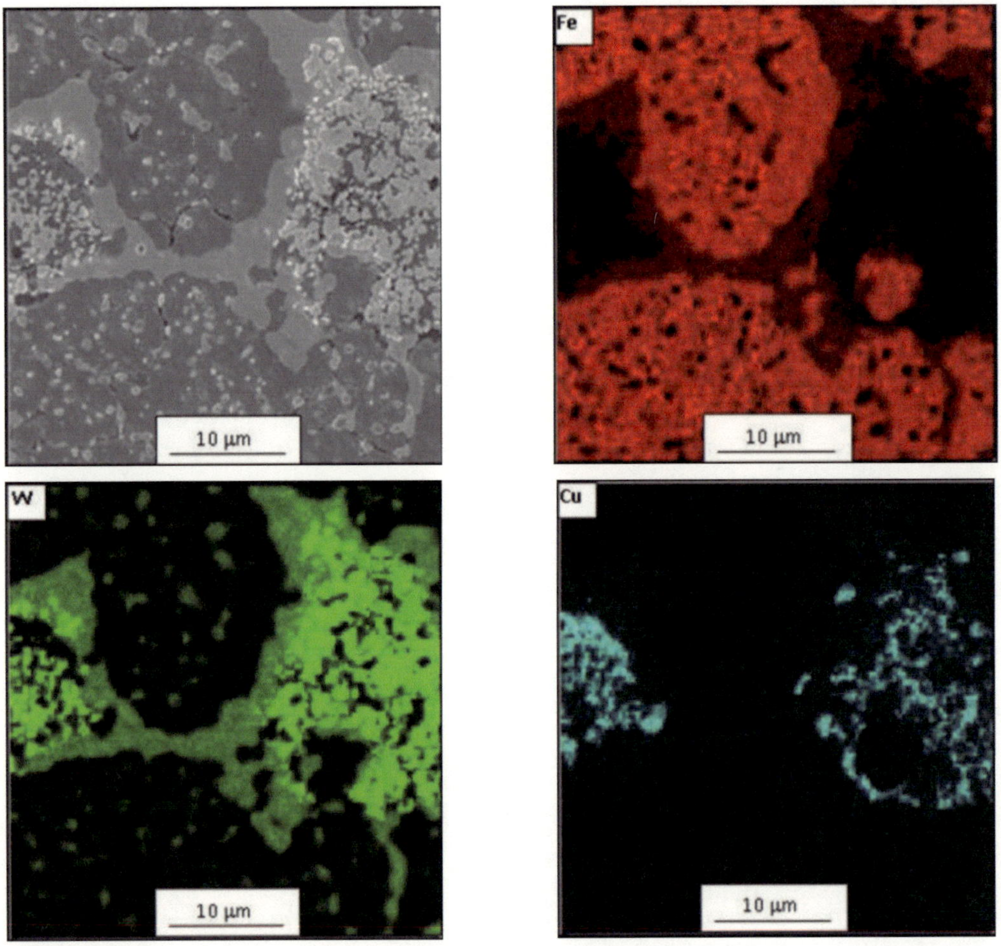

Figure 22. SEM micrograph and corresponding EDS maps of pre-sintered skeleton M3/2+30%WC infiltrated of copper

4.1. Tribological properties

All the specimens were polished to an average roughness of R_a = 1 µm. The tests were carried out at room temperature, keeping a relative humidity below 30%. The wear test results are given in Figures 23 and 24.

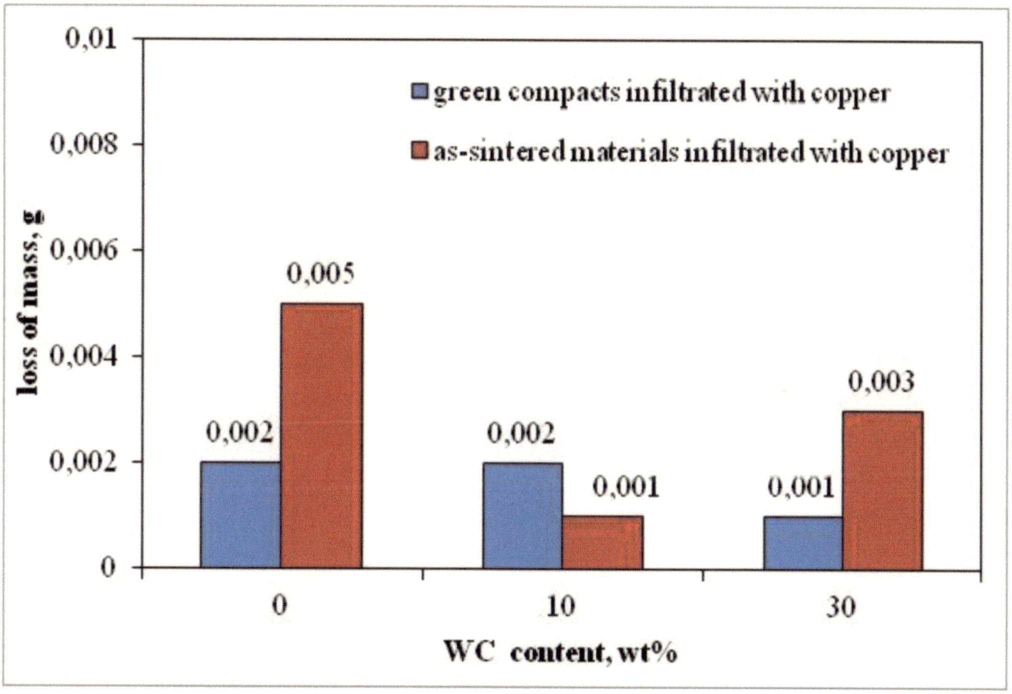

Figure 23. Loss of mass of as infiltrated composites

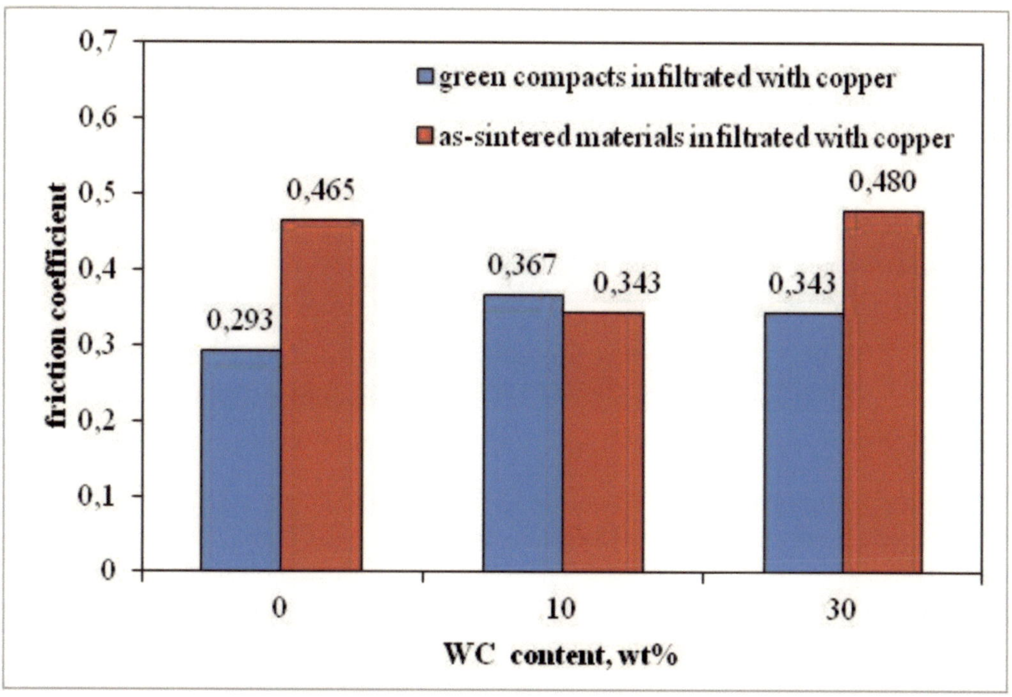

Figure 24. Friction coefficient of as infiltrated composites

The measurements of the wear resistance and friction coefficient permit classification of the as-infiltrated composites with respect to their tribological properties. Direct infiltration of green compacts with copper results in the highest wear resistance and almost the same friction coefficient of the as-infiltrated M3/2 and M3/2+30% WC composites. By comparing the wear resistance of composites received through direct infiltration of green compacts and infiltration of pre-sintered skeletons it is evident that the green compacts M3/2 and M3/2+30%WC compositions show 2÷3 times higher loss of mass than the tungsten carbide containing as-sintered materials infiltrated with copper. This can be explained by the diffusion of carbon and alloying of iron particles during sintering and chemical reaction between the tungsten monocarbide and HSS matrix. Friction coefficients are not highly influenced by the tungsten carbide additions, but the additions of 30% WC to high speed steel and infiltration with copper increase the wear resistance of these composites comparing to base material (M3/2 HSS infiltrated with copper). Wear tracks were analyzed by SEM to clarify wear mechanisms. Characteristic surface topographies after the wear test are exemplified in Figures 25 and 26.

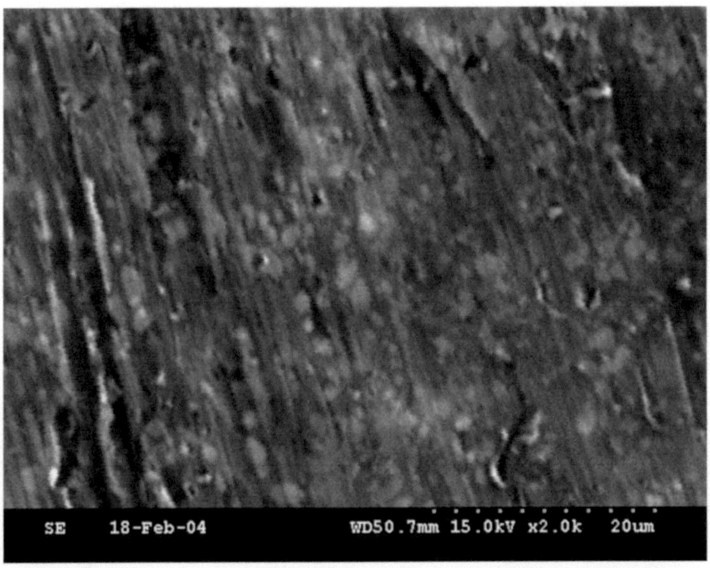

Figure 25. The surface of the as-infiltrated M3/2 composites after examining the wear resistance

The surface topographies of M3/2 and M3/2+30%WC specimens indicate occurrence of different wear mechanisms (Figure 25 and 26). In Fig. 25, typical abrasion scratches are seen in the base material. As a result of abrasion, ferrous oxides are generated and then dispersed through the wear track. The carbides seen on the wear-surfaces are being crushed and pulled out of the matrix to act as abrasive particles which increase the coefficient of friction. Figure 25 provide evidence of ploughing and sideways displacement of material in M3/2. Figure 26 shows smearing of iron oxides over the surface of the as-infiltrated M3/2+30%WC composite which implies marked contribution of adhesive wear, whereas the extensive formation of iron oxides may account for the higher friction coefficients. In MMCs, the level of oxidation is lower than in plain steel. The best behaviour is observed for composite green compact M3/2+30%WC infiltrated of copper. These carbides are well linked to the matrix and cannot be easily detached. M_6C formed as a result of chemical reaction between additions of WC and steel matrix are affected by abrasion, while MC carbides remain in the matrix withstanding wear and creating barriers where oxides from matrix debris are accumulated. WC forms large size agglomerates of small particles, which are detached when abraded and spread across the wear track. At first, these particles act as abrasives, promoting three-body wear behavior. The infiltration of green compacts has two effects: on one hand, fewer particles are detached from the carbide clusters, and on the other hand, these particles are not encrusted in the matrix, producing three-body abrasion in all compositions.

Figure 26. The surface of the as-infiltrated M3/2+30%WC composites after examining the wear resistance

The best materials from the viewpoint of mechanical properties were tested for wear properties is green compact M3/2 +30% tungsten carbide WC infiltrated with copper. Figures 23 and 24 suggest that, from the viewpoint of wear behaviour, the 30% reinforcement is the optimal composition for this type of composites.

5. Conclusions

1. Infiltration of porous HSS skeleton with liquid copper has proved to be a suitable technique whereby fully dense HSS based materials are produced at low cost.

2. Direct infiltration of green compacts with copper results in the higher hardness and higher resistance to wear of the M3/2 and M3/2+30 %WC composites, and allows to cut the production cost.
3. The mechanical properties of the HSS based composites are strongly dependent on the tungsten carbide content. The additions of tungsten carbide increase the hardness of HSS based composites, but decrease their bending strength.
4. Tungsten-rich M_6C type carbide is formed as a result of the chemical reaction between the tungsten monocarbide and HSS matrix
5. The carbides seen on the wear-surfaces of as infiltrated composites are being crushed and pulled out of the matrix to act as abrasive particles.

Author details

Marcin Madej
AGH University of Science and Technology,
Faculty of Metal Engineering and Industrial Computer Science, Krakow, Poland

Acknowledgement

This work was financed by Ministry of Science and Higher Education through the project No 11.11.110.788.

6. References

[1] Greetham G.: Development and performance on infiltrated and non-infiltrated valve seat insert materials and their performance. Powder Metallurgy, 1990, vol 3, no 2, pp. 112-114.
[2] Rodrigo H. Plama: Tempering response of copper alloy-infiltrated T15 high-speed steel, The International Journal of Powder Metallurgy, 2001, Vol. 37, No 5, s. 29-35.
[3] Wright C.S.: The production and application of PM high-speed steels. Powder Metallurgy 1994 vol 3, pp. 937-944.
[4] Torralba J.M. G. Cambronero, J. M. Ruiz-Pietro, M. M das Neves: Sinterability study of PM M2 and T15 HSS reinforced with tungsten and titanium carbides, 1993, vol 36, pp. 55 – 66.
[5] M. Madej, J. Leżański: Copper infiltrated high speed steel based composites, Archives of Metallurgy and Materials, 2005 vol. 50 iss. 4 s. 871–877.
[6] M. Madej, J. Leżański: The structure and properties of copper infiltrated HSS based, Archives of Metallurgy and Materials, 2008, vol. 53, iss. 3 s. 839–845.
[7] M. Madej: The tribological properties of high speed steel based composites, Archives of Metallurgy and Materials 2010 vol. 55 iss. 1 s. 61–68

[8] L.A. Dobrzanski, [..]: Fabrication methods and heat treatment conditions effect on tribological properties of high speed steels, Journal of Metarials Processing Technology, 157-158, (2004), s. 324-330.

[9] E. Gordo, F. Velasco, N. Anto´n, J.M. Torralba: Wear mechanisms in high speed steel reinforced with (NbC)p and (TaC)p MMCs, Wear 239 (2000), s. 251–259

[10] Farid Akhtar: Microstructure evolution and wear properties of in situ synthesized TiB2 and TiC reinforced steel matrix composites, Journal of Alloys and Compounds, 459 (2008), s. 491–497

[11] G. Hoyle: *High Speed Steels.* Butterworth & Co. Publishers. Cambridge 1998

[12] Shizhong Wei, Jinhua Zhu, Liujie Xu: Effects of vanadium and carbon on microstructures and abrasive wear resistance of high speed steel, Tribology International 39 (2006), s. 641–648

[13] Z. Zalisz, A. Watts, S.C. Mitchell, A.S. Wronski: Friction and wear of lubricated M3 Class 2 sintered high speed steelwith and without TiC and MnS additives, Wear 258 (2005), s. 701–711

[14] W. C. Zapata, C. E. Da Costa, J. M. Torralba: Wear and thermal behaviour of M2 high-speed steel reinforced with NbC composite, Journal of Materials science, 33 (1998) 3219 – 3225

[15] G. A. Baglyuk and L. A. Poznyak: The sintering of powder metallurgy high-speed steel with activating additions, Powder Metallurgy and Metal Ceramics, Vol. 41, No 7-8, 2002, s. 366-368

[16] W. Khraisat, L. Nyborg and P. Sotkovszki: Effect of silicon, vanadium and nickel on microstructure of liquid phase sintered M3/2 grade high speed steel, Powder Metallurgy 2005 Vol. 48 No. 1 s. 33-38

[17] J. A. Jime´nez, M. Carsı, G. Frommeyer and O. A. Ruano: Microstructural and mechanical characterisation of composite materials consisting of M3/2 high speed steel reinforced with niobium carbides, Powder Metallurgy 2005 Vol. 48 No. 4, s. 371-376

[18] J. D. Bolton and A. J. Gant: Phase reactions and chemical stability of ceramic carbide and solid lubricant particulate additions within sintered high speed steel matrix, Powder Metallurgy 1993 Vol. 36 No.4, s. 267-274.

[19] J. D. Bolton and A. J. Gant: Heat treatment response of sintered M3/2 high speed steel composites containing additions of manganese sulphide, niobium carbide, and titanium carbide, Powder Metallurgy 1996 Vol. 39 No.1, s. 27-34.

[20] M. Madej: Copper infiltrated high speed steel based composites with iron additions, Archives of Metallurgy and Materials, 2009 vol. 54 iss. 4 s. 1083–1091

[21] M. Madej: The tribological properties of high speed steel based composites, Archives of Metallurgy and Materials, 2010 vol. 55 iss. 1 s. 61–68

[22] H. G. Rutz and F. G. Hanejko: High density processing of high performance ferrous materials, international conference & Exhibition on powder Metallurgy & Particulate Materials, May 8-11, 1994 - Toronto, Canada

[23] M. M. Oliveira: *High-speed steels and high-speed steels based composites*. International Journal of Materials and Product Technology, 2000, Vol. 15, No 3÷5, s. 231-251.

Chapter 4

Tungsten Carbide as an Reinforcement in Structural Oxide-Matrix Composites

Zbigniew Pędzich

Additional information is available at the end of the chapter

http://dx.doi.org/10.5772/51183

1. Introduction

The possibility of serious improvement of mechanical properties of oxide ceramics by particulate composites manufacturing has been recently recognized very well. Among oxide ceramics, tetragonal zirconia solid solutions and α-alumina phase are the most important materials, widely used in structural applications, due to their good properties. The fabrication of two-phase particulate composites could be the simplest way to the mechanical properties improvement. Despite a wide range of alumina-zirconia composites, non-oxide particles were also often utilized as strengthening agents. Many phases were introduced into zirconia and alumina matrices – TiC, SiC, WC, TiB_2, TiN, AlN, (Ti,W)C, Cr_2O_3, Cr_7C_3, and metals – nickel, molybdenum and tungsten and others [1-17]. In this way, the materials with improved properties, when compared with "pure" matrix materials, were obtained. Depending on the type of inclusions, their size and amount as well as sintering conditions, one can achieve a significant improvement of hardness, stiffness, fracture toughness and/or strength of the material. It was also reported that the decrease of inclusion size to the nanometric scale allowed extremely high values of flexural strength and fracture toughness to be achieved.

The manufacturing of composites with ceramic matrix almost always leads to residual stresses caused by the mismatch of thermal properties of constituent phases. A large difference in thermal expansion coefficients (CTE's) could introduce stresses reaching even more than gigaPascals to the composite system. Such a phenomenon has to influence the way of fracture and consequently the strength and the fracture toughness of the material. The value of these stresses mainly depends on mechanical properties of constituent phases of the composite and the absolute difference between their CTE's. The distribution of residual stresses depends also on the phase arrangement and shape of grains. This chapter presents the investigation results on the influence of the phase arrangement on the way of fracture in composites. These observations were put together with the results of mechanical properties measurements and abrasive wear tests.

2. Thermodynamical aspect

2.1. ZrO_2/WC system

Zirconia for structural applications is used in the form of solid solutions of yttria, magnesia, calcia or rare earth metals in ZrO_2 [18-20]. Using data from [21] one can calculate that free enthalpy of mixing of zirconia and any stabilizing element is significantly lower than the error of determination of free enthalpy for chemical reactions in ZrO_2-WC (or ZrO_2-WC-C) systems. It allows to calculate, with a reasonable approximation, the possibility of reactions proceeding using thermo-dynamical data for zirconia only.

Potential chemical reactions taken into account were as follow:

$$ZrO_2 + WC = ZrC + WO_2 \quad (1)$$

$$ZrO_2 + 6WC = ZrC + 3W_2C + CO \quad (2)$$

Calculations showed that reaction (1) cannot proceed in the range of potential sintering temperatures (1400 - 1700°C) because of fact that standard free enthalpy (ΔG_r^0) of that reaction is much higher than zero.

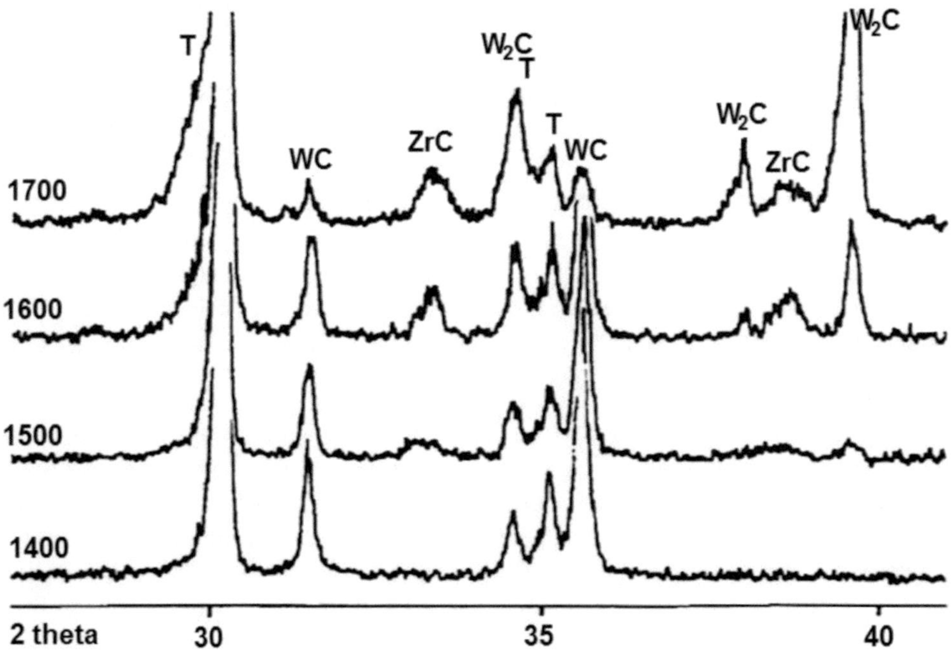

Figure 1. Results of XRD analyses of 3Y-ZrO_2/10vol.% of WC composite pressureless sintered at different temperatures. T: stand for tetragonal phase of the zirconia solid solution, sintering temperatures indicated on the left side of plots.

Reaction (2) can proceed when the partial pressure of CO in the system is lower than 0.95 atm at 1400°C, 5.20 atm at 1500°C, 23.8 atm at 1700°C and 93.0 atm at 1700°C.

These calculations were verified experimentally [22]. The presence of ZrC and W_2C was determined in sinters containing WC inclusions sintered at different temperatures (Fig. 1.).

If sintering process is conducted using hot-pressing (HP) technique, composite powder could be in contact with carbon from the press die or stamps. It suggest that the third reaction should be also taken into account:

$$ZrO_2 + 3C = ZrC + 2CO \tag{3}$$

Reaction (3) can proceed when the partial pressure of CO in the system is lower than 0.054 atm at 1400°C, 0.20 atm at 1500°C, 0.65 atm at 1700°C and 1.89 atm at 1700°C. These data suggest that is possible to produce ZrC precipitates even in pure zirconia sinters when one can assure right value of CO pressure [17]. Conducting of sintering process at relatively low temperature (1400°C) allows to avoid ZrC appearance (Fig. 2).

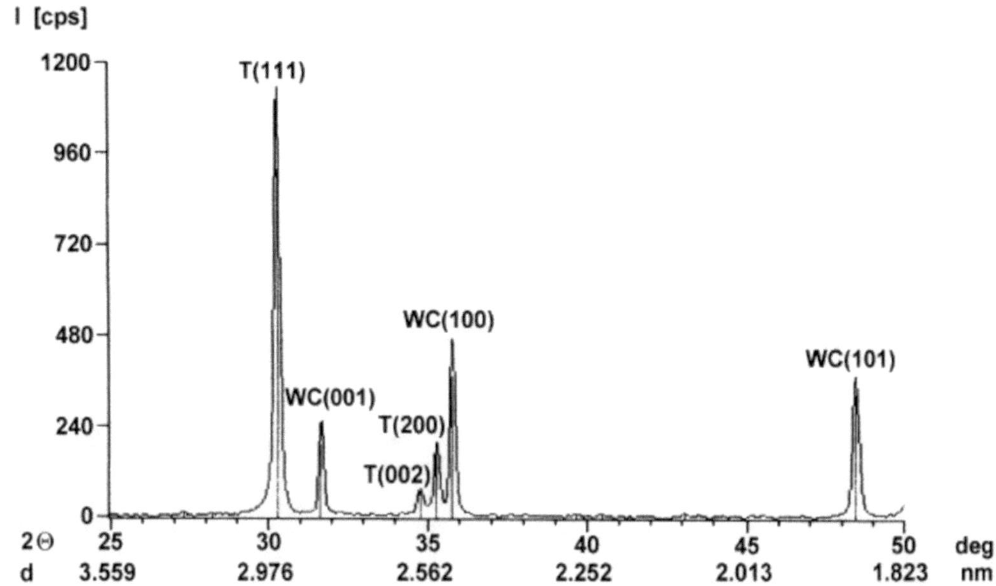

Figure 2. Result of XRD analysis of 3Y-ZrO_2/10vol.% of WC composite hot-pressed at 1500°C. T: stand for tetragonal phase of the zirconia solid solution.

2.2. Al_2O_3/WC system

Potential chemical reactions in alumina – tungsten carbide systems taken into account:

$$2Al_2O_3 + 3WC = Al_4C_3 + 3WO_2 \tag{4}$$

$$2Al_2O_3 + 6WC = Al_4C_3 + 3W_2C + 3CO_2 \tag{5}$$

$$2Al_2O_3 + 9WC = Al_4C_3 + 9W + 6CO \quad (6)$$

Calculations showed that reactions (4 - 6) cannot proceed in the range of potential sintering temperatures (1400 - 1700°C) because of fact that standard free enthalpy (ΔG_r^0) of that reaction is much higher than zero. That results were also confirm by Niyomwas [23], who stated that Al_2O_3/WC system is thermodynamically stable up to 2000°C.

3. Internal stress state

The manufacturing of composites with ceramic matrix almost always leads to residual stresses caused by the mismatch of thermal properties of constituent phases. The value of these stresses mainly depends on mechanical properties of constituent phases of the composite and the absolute difference between their CTE's. A large difference in thermal expansion coefficients (CTE's) could introduce to the composite system stresses reaching even more than gigaPascals. Such a phenomenon has to influence mechanical properties of the material. The distribution of residual stresses depends also on the phase arrangement and shape of grains.

Phase	CTE (α), $\cdot 10^{-6}C^{-1}$	Young modulus E, GPa	Poisson ratio, ν
Alumina	7.9	385	0.250
Zirconia ss.	11.0	210	0.210
WC	5.2	700	0.300

Table 1. Data necessary for calculation of the residual stresses value.

The thermal expansion coefficient of tungsten carbide (α_{WC}) is lower than thermal expansion coefficients of both considered oxide phases (α_{Al2O3} and α_{ZrO2}). It means that in composites with both oxide matrices the internal stress state is similar. Matrices are under tension and carbide inclusions are under compression.

For this chapter results of calculation of stresses in materials was made using the finite elements model (FEM) based on following predictions:

- the grain in the matrix in two-dimensional geometry,
- the model was constrained to enable a free deformation in *xy* plane to be carried out and, additionally, in one corner,
- the geometric model was discretized with the AutoGEM modulus [24, 25]. For calculations the elements neighboring the point of support were excluded. This eliminated the stress accumulation at the model edge,
- grain boundaries inside constituent phases were omitted,
- calculations were made using the mechanical property values (Young's moduli, Poisson ratio's and CTE's) placed in Table 1. Isotropy of these constants was taken as a principle,
- modeling was performed for the plain stress state,
- Used method of load was cooling from temperature of 1200°C to room temperature

The results of FEM simulations were visualized at Figures 3 and 4. They present the distribution of principal maximal stresses around in the hypothetical composite microstructure. Calculations were made for the same schematic microstructure. Matrix was assumed as zirconia or alumina, respectively. The inclusion phase was WC. Calculations were made for Al_2O_3/WC and ZrO_2/WC composites.

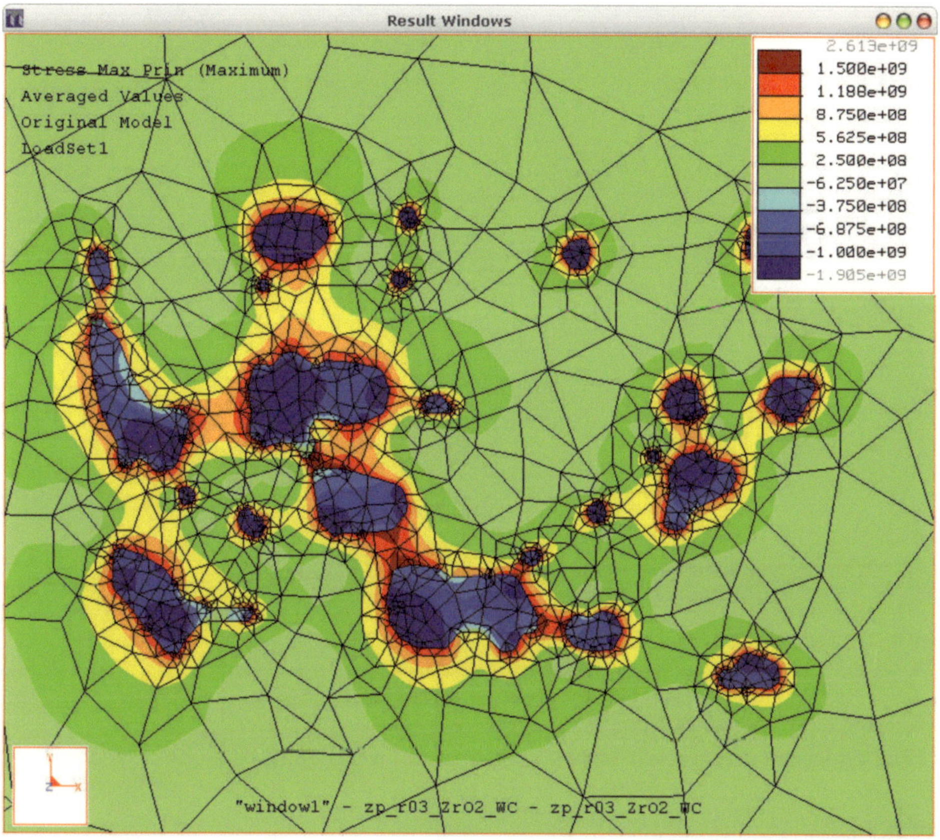

Figure 3. The principal maximal stresses calculated for ZrO_2/WC composite. Dark blue color represents the maximal values of compressive stresses, brown color represents the maximal values of tensile stresses. At this Figure WC inclusions are generally in blue color.

Generally, the maximum value of principal maximal stresses in the zirconia matrix is about 30 % higher than in the alumina one. The tensile stress level near the interphase boundary in the zirconia matrix materials exceeds 1000 MPa all around the inclusion grain (Fig. 3). In the alumina based materials maximum stress values in this area are much lower (Fig. 4).

This fact influences the path of crack in the investigated materials. In zirconia-based composites crack goes along the interphase boundary (Fig. 5). The crack course in composites with alumina matrix is different. It usually goes near the inclusion grains, but it is deflected before it reaches the interphase boundary (Fig. 6). This means that the crack goes through alumina grains.

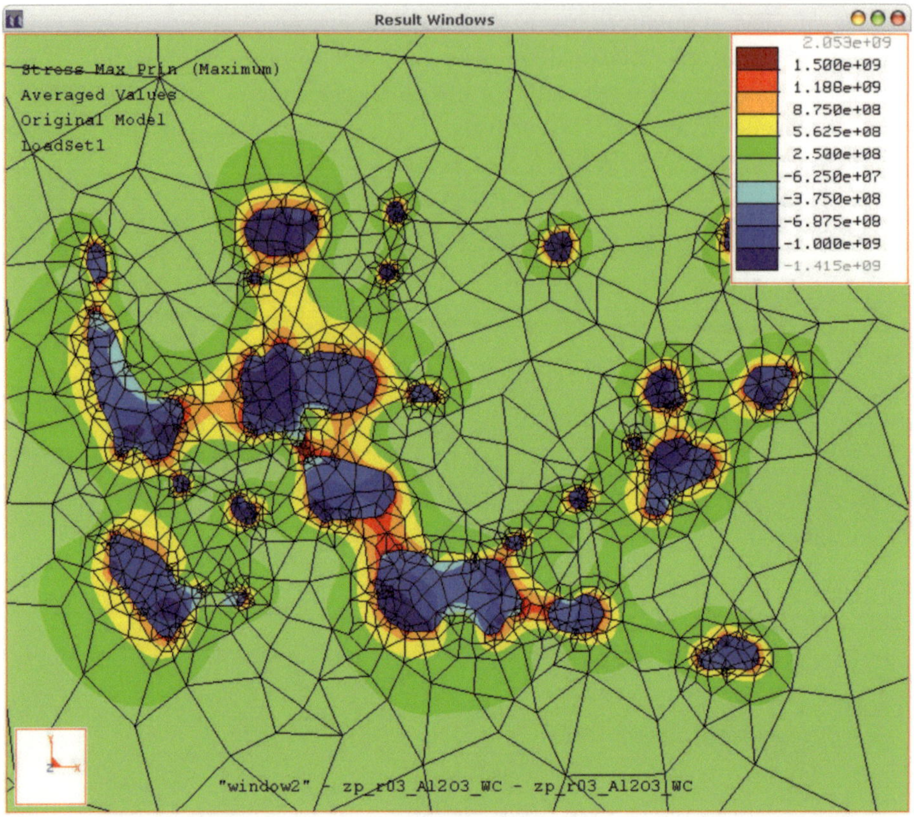

Figure 4. The principal maximal stresses calculated for Al_2O_3/WC composite. Dark blue color represents the maximal values of compressive stresses, brown color represents the maximal values of tensile stresses. At this Figure WC inclusions are generally in blue color.

The final effect of such crack behaviour for material toughening is summarized in Table 4. As it is clearly visible, the relative fracture toughness increase observed for the alumina-based composites is higher than for the zirconia ones.

This phenomenon should be attributed to the lower stress level in the alumina-based composites. As it can be seen at Figures 3 – 4, the maximum stress values are present in some distance between inclusion grains. Probably the strength of alumina grain is comparable with the strength of interphase boundaries (Al_2O_3/WC) in composites. Such a situation promotes transgranular cracking of alumina (see Fig. 6), but in a specific way, the crack still wanders around inclusions and crack deflection mechanism is still active and it consumes energy effectively.

In TZP matrix composites the tensile stress acting on the interphase boundary is much higher than in these with alumina matrix. It decrease the amount of energy dissipated during cracking. Additionally, high toughness of the zirconia material causes that the crack does not deflect as in the case of alumina. The crack rather goes to the interphase boundary and deflects directly on it. These observations are only qualitative but they could help to understand the effect of a relatively high level of toughening in the alumina based composites.

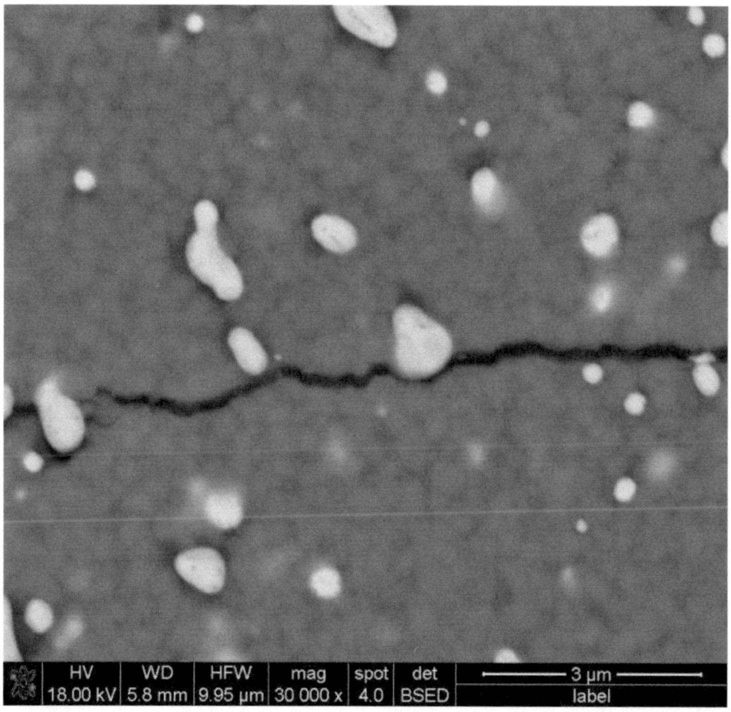

Figure 5. The SEM image of crack path in ZrO_2/WC composite.

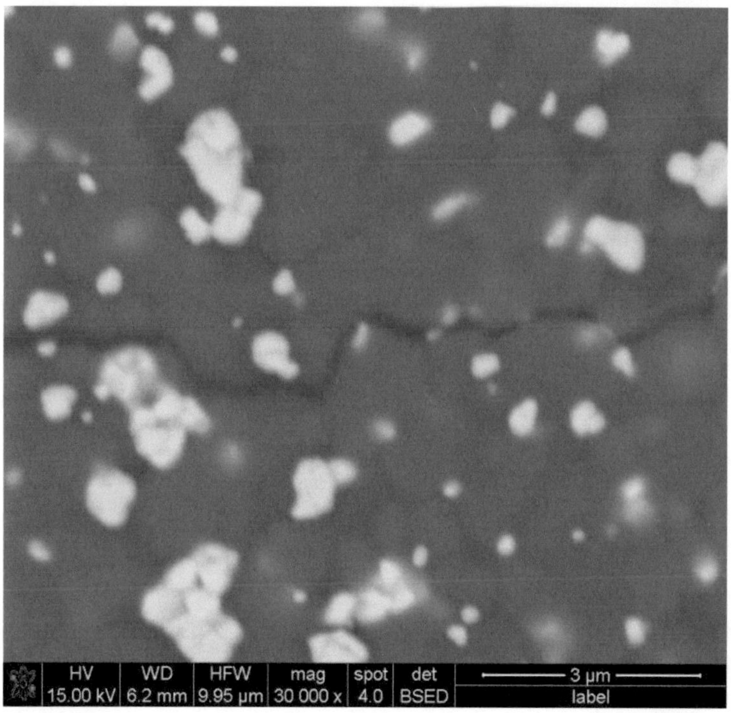

Figure 6. The SEM image of crack path in Al_2O_3/WC composite.

4. Composite manufacturing

Preparation of dense particulate composite bodies with randomly distributed inclusions meets potential difficulties during composite powder preparation and during sintering. The most popular method of second phase dispersion in the matrix is simple mixing. This method is widely used for zirconia/WC system [26-28]. The mixing process utilizing intensive mills (atrittors, rotation-vibrational mills) in short time assures the proper tungsten carbide particles distribution within the matrix in the wide range of WC content 10 – 50 vol. %. More sophisticated methods as for instance decomposition of organic WC precursors are nowadays too expensive for wider application [23, 29, 30].

In alumina/WC composite system the mechanical mixing is also the main preparation method of the composite powder. Anyway, there were some experiments [23] utilizing self-propagating high temperature synthesis (SHS) process for *in-situ* synthesis of alumina/tungsten carbide composite powder. In this process both tungsten carbides (WC, W_2C) were present in the product.

Sintering of composites with oxide matrices and dispersed WC particles is a typical example of sintering with "rigid inclusions" widely described in literature [31-33]. In fact, during this process, diffusional mechanisms of densifications appear only in the oxide matrix. The presence of carbide particles makes the sintering driving forces much weaker. This effect is as stronger as higher is tungsten carbide particles volume content. The high relative density demand for structural applications (> 97 % of theo.) can be assured using pressureless sintering method when WC content not exceed 20 vol. %. Additionally, sintering temperature in this case must be relatively high (1550°C for zirconia and 1600°C for alumina). It is not profitable for sinters microstructure because the grain growth phenomenon. The inert sintering atmosphere demanded for preservation of WC from oxidation at high temperatures causes some additional factor of stabilization in zirconia [22, 34].

Practically, the most often sintering method for both type of composites is hot-pressing technique. Application of this method allows to assure high densities (> 98 % of theo.) in relatively short time (30 – 60 min.). Such conditions limits the grain growth in the matrix (see Table 3).

There were some investigations utilizing pulsed electric current sintering (PECS) for zirconia/WC composite densification [27]. These methods were profitable when WC content in the composite was relatively high (~30-40 vol. %).

5. Composites microstructure

In present chapter author was focused on properties of "classical" particulate composites. It means materials containing the second phase particles randomly distributed into matrix. It means that the amount of additives must be lower than a percolation threshold. To assure that situation examples of composite materials contain 10 vol. % of tungsten phase with the same grain size distribution were manufactured.

Commercial powders were used as a starting materials (alumina – TM-DAR Taimicron, zirconia – Tosoh 3Y-TZ, tungsten carbide – Baildonit). Powders homogeneity was assured by 30 min. of attrition mixing of constituent powders in ethyl alcohol.

Materials for test were fabricated by hot-pressing technique (HP) due to guarantee the maximum of the densification of investigated samples. The sintering conditions were as follow: the maximum temperature - 1500°C (for zirconia and zirconia/WC composite) and 1650°C (for alumina and alumina/WC composite) with 1 hour soaking time and maximum applied pressure - 25 MPa.

A typical SEM microstructures of hot-pressed composites were showed in the Figs. 7 and 8. Table 2 collects data about the grain size of individual phases.

Material	Mean grain size, μm		
	Al_2O_3	ZrO_2	WC
Alumina	5.20 ±2.90	-	-
Alumina/10vol.% WC	1.25 ±0.80	-	0.45 ±0.30
Zirconia solid solution	-	0.32 ±0.18	-
Zirconia/10vol.% WC	-	0.27 ±0.15	0.47 ±0.35

Table 2. The mean grain size of phases existing in sinters containing 10 vol. % of WC.

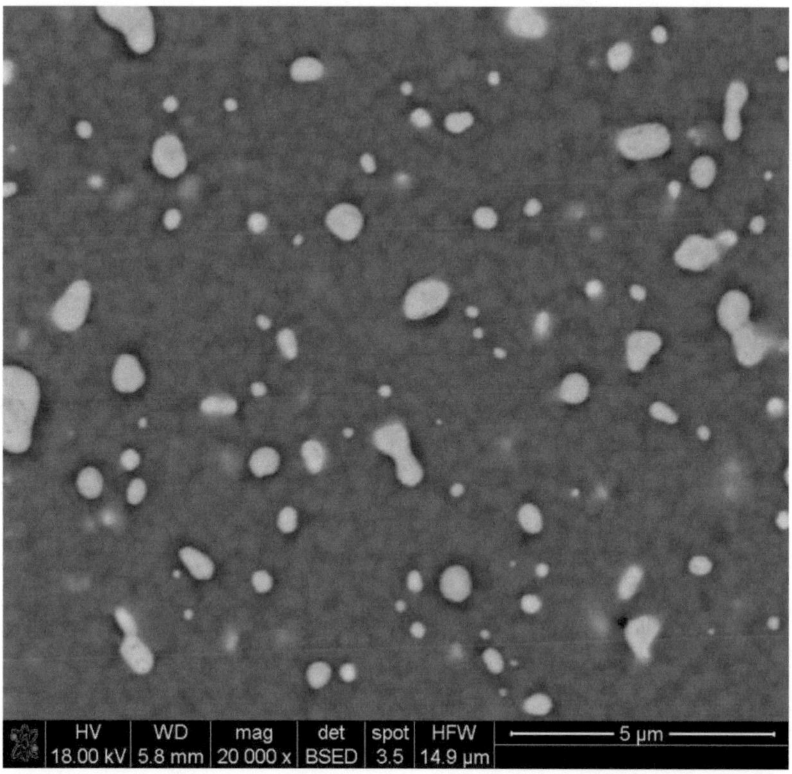

Figure 7. The typical SEM image of thermally etched ZrO_2/WC composite microstructure.

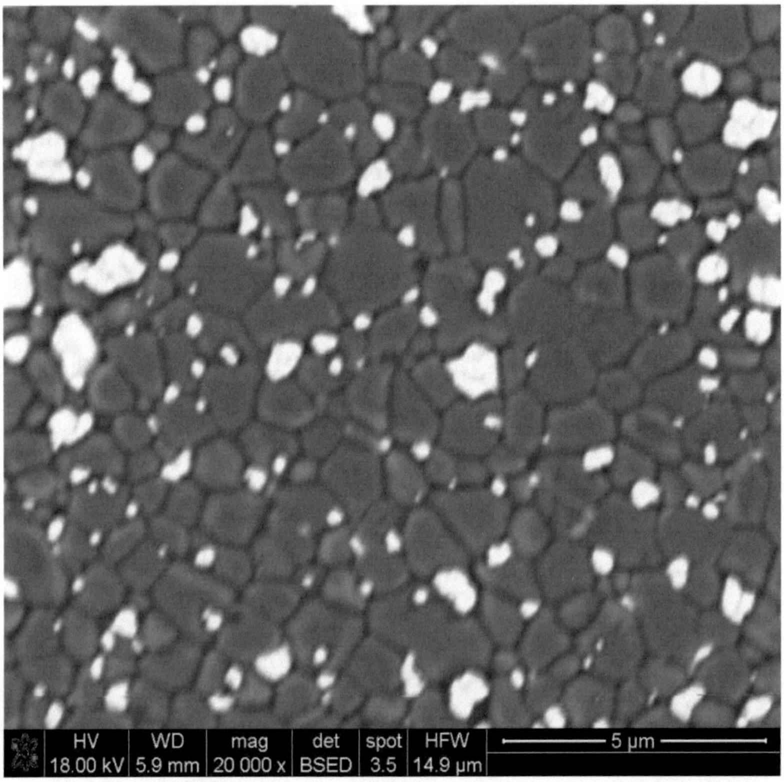

Figure 8. The typical SEM image of thermally etched Al_2O_3/WC composite microstructure.

Measurements performed in the TEM revealed that oxide matrices and tungsten carbide grains close adhered and no discontinuities were observed (Figs. 9 and 10).

The detail observation of Al_2O_3/WC and ZrO_2/WC microstructures and chemical analyses performed as a line scan across the interphase boundaries (Figs.11 and 12) showed that there are differences in elements diffusion in investigated systems. The change of chemical composition near alumina/tungsten carbide boundary is sharp and distinct (Fig. 9). There is no evidence of Al diffusion into WC or W diffusion into Al_2O_3. In the case of zirconia/tungsten carbide boundary chemical composition is changing near the interphase boundary. It could be a slight confirmation of thermo-dynamically described tendency to creation of ZrC and W_2C in this system.

Results of TEM investigations have shown specific crystallographic relationships between alumina and zirconia matrix and tungsten carbide phase [35]. Crystal correlations may partially explain significant improvement of mechanical properties of alumina- and zirconia-based composites comparing with a pure oxide matrices. However, apart from crystallographic factors, the properties of material under investigation may be affected by interfacial defects and interphase boundary structure.

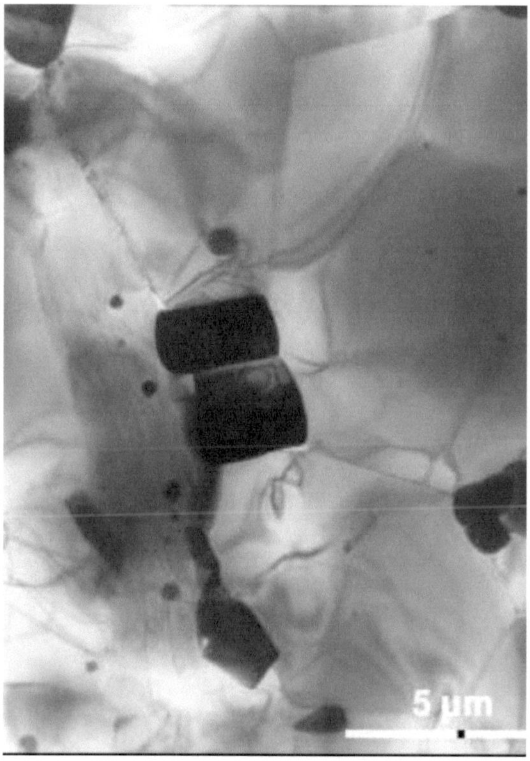

Figure 9. TEM micrograph of Al_2O_3/WC composite. Dark grains – WC; bright ones – alumina.

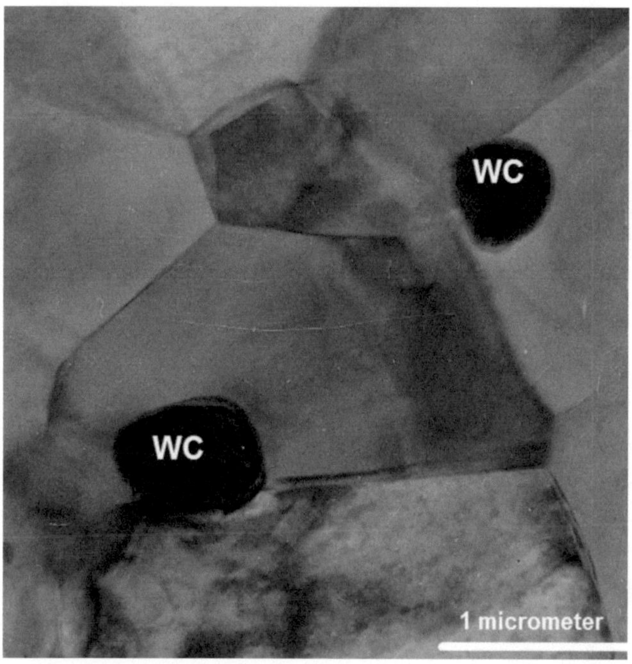

Figure 10. TEM micrograph of ZrO_2/WC composite.

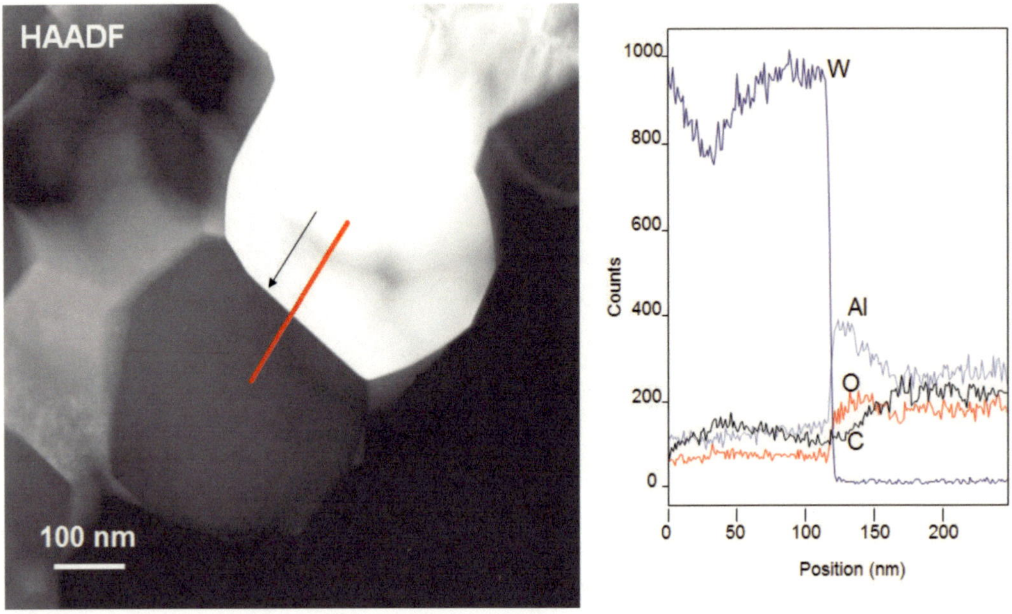

Figure 11. High resolution TEM microstructure of Al_2O_3/WC and the line scan across alumina and tungsten carbide boundary.

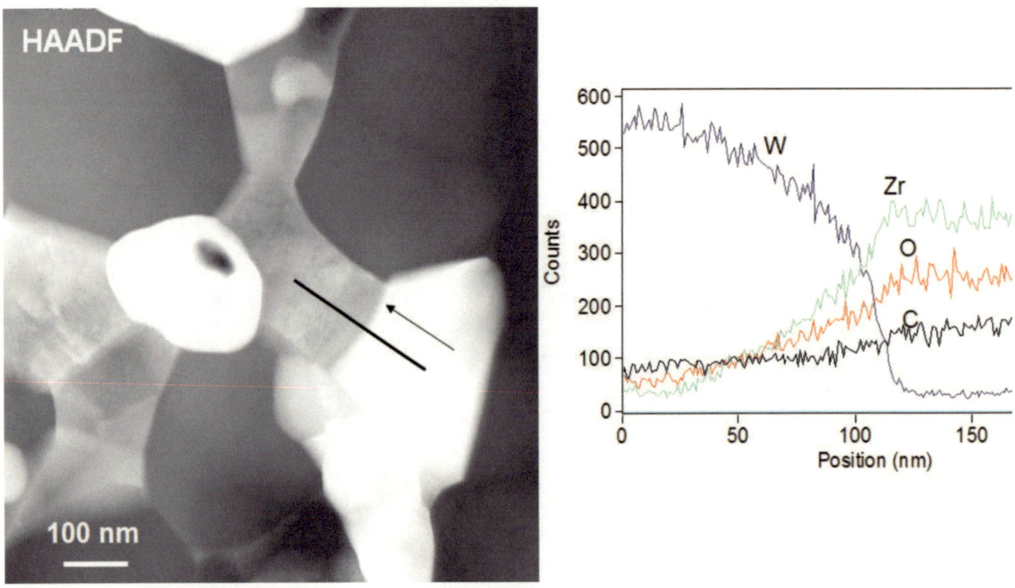

Figure 12. High resolution TEM microstructure of ZrO_2/WC and the line scan across zirconia and tungsten carbide boundary.

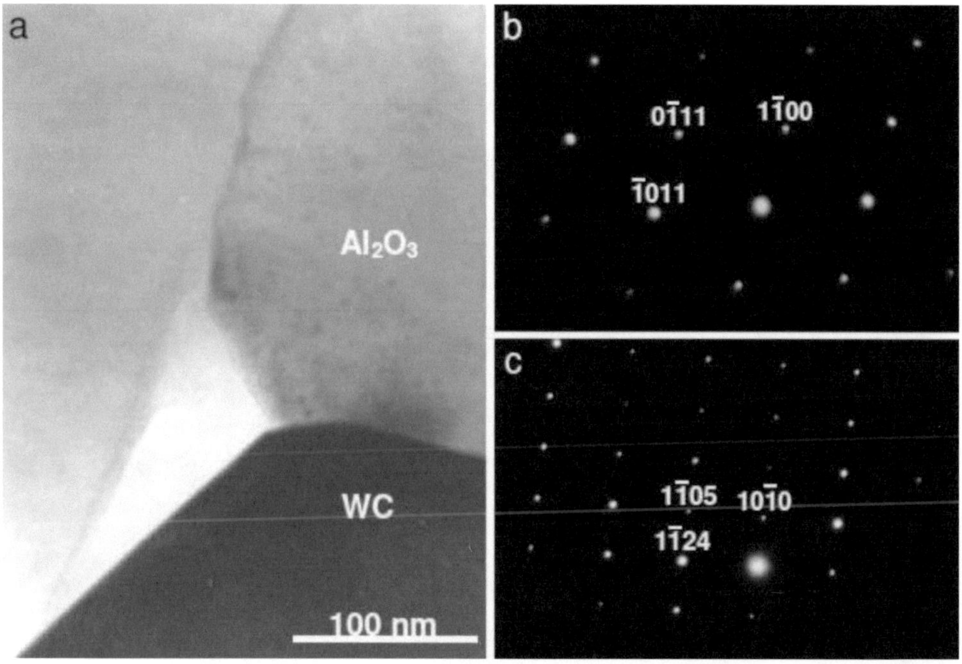

Figure 13. TEM micrograph of Al_2O_3/WC composite. a – bright field (BF) image, b – selected area electron diffraction (SEAD) from WC grain, c - SEAD from Al_2O_3 grain.

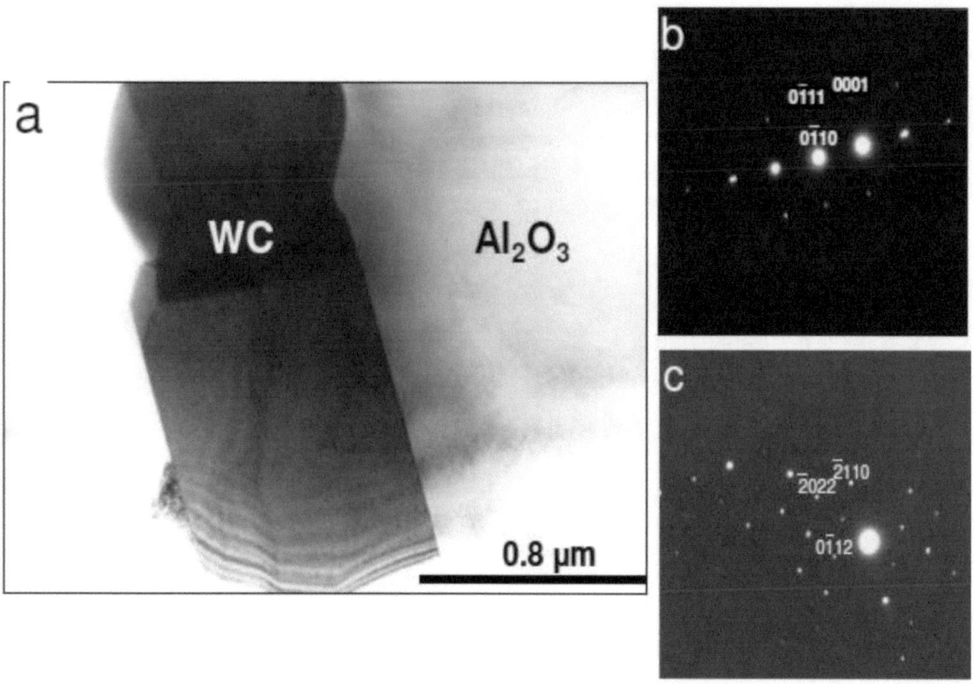

Figure 14. TEM micrograph of Al_2O_3/WC composite. a – BF image, b –SEAD from WC grain, c - SEAD from Al_2O_3 grain.

Alumina and WC grains were indexed using the SAED "(Selected Area Electron Diffraction) and two characteristic crystal relationships between above phases were identified (Figs. 13 and 14):

$$(0\ 111)\ WC \parallel (1\ 105)\ Al_2O_3 \tag{7}$$

$$[11\ 23]\ WC \parallel [23\ 11]\ Al_2O_3 \tag{8}$$

and

$$(0\ 111)\ WC \parallel (1011)\ Al_2O_3 \tag{9}$$

$$[2\ 1\ 10]\ WC \parallel [01\ 11]\ Al_2O_3 \tag{10}$$

These relationships were found on several sites investigated on the thin foil.

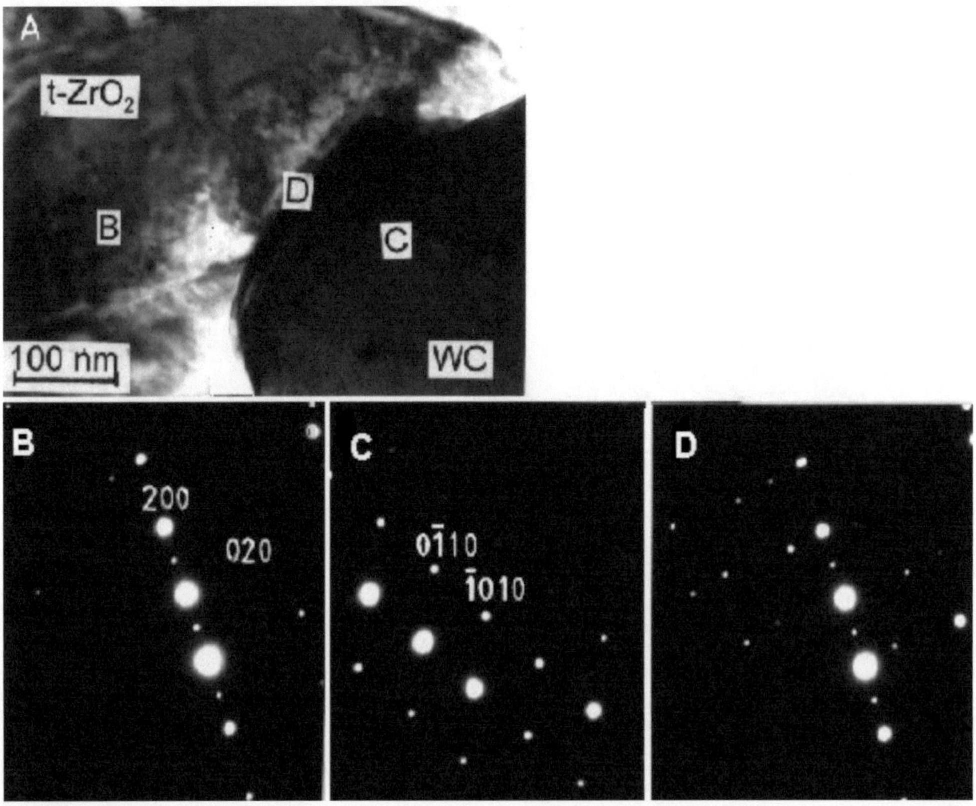

Figure 15. TEM micrograph of ZrO_2/WC composite. A – BF image, B –SEAD from ZrO_2 grain, C– SEAD from WC grain, D – SEAD from the grain boundary region.

Similarly, in ZrO_2/WC system crystallographic relationships were identified (Fig. 15) [36]:

$$[0001]\ WC \parallel [001]\ \text{tetragonal}\ ZrO_2 \tag{11}$$

$$[\bar{1}010]\ \text{WC} \parallel [010]\ \text{tetragonal}\ ZrO_2 \qquad (12)$$

Furthermore, EBSD analysis made by Faryna at. all. [37, 38] proved statistically that crystallographic correlation in investigated composite systems are not an unique property, but they are very often.

6. Mechanical properties

The basic mechanical properties of investigated materials were collected in Table 3. Both composites were well densified but is worth to noticed that there is about 1 % of difference between Al_2O_3/WC and ZrO_2/WC composites. Zirconia matrix and zirconia-basing material is almost fully dense. Alumina-basing composite has "only" 98.8 % of theoretical density and is 0.5 % worse densified than alumina matrix. This difference is not much but certainly influence observed bending strength test results.

It is characteristic that Al_2O_3/WC material has lower bending strength than "pure" matrix material. Different effect is observed for ZrO_2/WC composite. The mean value of the bending strength of ZrO_2/WC is similar to that registered for zirconia matrix. But the highest strength value registered during tests was over 10% higher than that measured for zirconia matrix. This fact showed that there is a potential of strength improvement in this system.

It is not surprise that hardness of composites is higher than that measured for matrices. Spectacular is the increase of the fracture toughness. In both investigated composite systems K_{Ic} increased more than 50 % when compared with the suitable matrix.

Material	Relative density, $\varrho_{wzgl.}$, %	Vickers hardness, *HV*, GPa	Young's modulus, *E*, GPa	Fracture toughness, K_{Ic}, MPam^{0,5}	Bending strength – the mean value, σ, MPa	Bending strength – maximum value, σ_{max}, MPa	Weibull parameter, *m*
Alumina	99.3 ± 0.1	17.0 ± 1.2	379 ± 6	3.6 ± 0.3	600 ± 120	780	6
Alumina/10vol.% WC	98.8 ± 0.1	18.7 ± 0.8	394 ± 7	5.5 ± 0.7	450 ± 45	550	12
Zirconia s.s.	99.5 ± 0.1	14.0 ± 0.5	209 ± 5	5.0 ± 0.5	1150 ± 75	1250	18
Zirconia/10vol.% WC	99.7 ± 0.1	17.0 ± 0.9	232 ± 6	8.0 ± 1.0	1100 ± 130	1380	7

± denotes the confidence interval on the 0.95 confidence level (for ϱ, *HV* and K_{Ic} measurements);

± denotes the standard deviation of the mean value of 40 results of measurements (for σ measurements).

Table 3. Mechanical properties of the matrices and composites.

Experiments of subcritical crack propagation performer using Double Torsion method (DT) [39- 41] showed that composites were much more resistant for this disadvantageous phenomenon. Such experiments were previously conducted for alumina and zirconia materials [42, 43] but not for composites containing tungsten carbide particles. Results of performed investigations (see Fig. 16) showed that the threshold value of K_I coefficient in both composite systems significantly increased. In Al_2O_3/WC material the threshold K_I value was ~4.0 $MPam^{-0.5}$ (compared with 2.6 $MPam^{-0.5}$ for alumina). In ZrO_2/WC material the threshold K_I value was ~4.4 $MPam^{-0.5}$ (compared with 3.6 $MPam^{-0.5}$ for zirconia). The most probably reason of such behaviour was the residual stresses state in dense sintered composite bodies. Distribution of these stresses around composite hindered breaking of atomic bonds on the tip of flaws presented in composites.

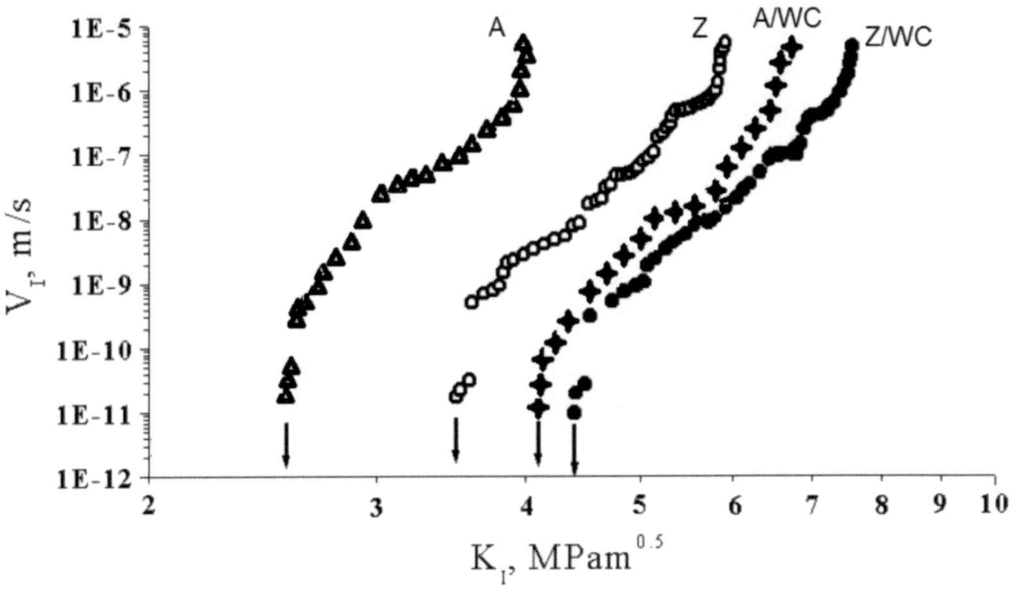

Figure 16. The crack velocity V_I vs. stress intensity factor K_I. A, Z, A/WC, Z/WC – stand for alumina, zirconia, alumina/WC composite and Zirconia/WC composite, relatively.

7. Wear resistance

One of the most important field of structural ceramic application is using them as a part of devices resistant for abrasive wear. From this point of view it is important how the material behaves at different working conditions (the range of loads) and environments (the presence of humidity). The oxide matrices are sensitive for water presence in environment. Even under low load rate it can cause the subcritical crack growth [42 - 45]. If the load are serious, degradation of the oxide matrices in the presence of water could be very significant.

The chapter presents the results of investigation on wear of alumina and zirconia-basing composites by hard abrasive particles in different environments. The results of two tests (The Dry Sand Test and the Miller Test in pulp) were compared. As an abrasive medium coarse silicon carbide grains were used in both cases.

The type of Dry Sand Test based on ASTM test [46], which indicates wear susceptibility of material for wear during abrasive action of hard particles without any lubricant. The test duration was 2000 rotation of the wheel.

The Miller Test based on ASTM test [47] allows to determine of SAR (Slurry Abrasion Response) number during the wear in slurry. The test duration was 6 hours. In both test the low value of test result means the better material behaviour.

Results of performed wear tests were collected in Table 4.

Material	Miller Test, SAR number,	Dry Sand Test, wear rate in mm^3
Alumina	163.51	55.17
Alumina/10vol.% WC	9.57	9.57
Zirconia s.s.	9.45	15.71
Zirconia/10vol.% WC	7.41	11.46

Table 4. Results of performed wear tests.

The results showed difference in wear mechanisms between tests conducted in dry and wet environments for composites with two matrices: α-alumina and tetragonal zirconia. Figures 17 and 18 collects SEM images of worn surfaces after the Dry Sand Test and the Miller Test. The fundamental difference visible at these figures, especially for alumina phase, is the presence of intensive grain boundary etching during work in wet environment. It could leads even to the whole grain scouring from the sintered bodies. During the Dry Sand Test the dominant wear mechanism consist in grains fracture. The presence of hard WC particles in both investigated oxide matrices limits the wear rates.

The worn surface profilographic analysis (see Table 5) allows to establish that second phase particle addition modifies alumina microstructure significantly. Alumina-basing composite surface after both tests were much more smooth, than the pure matrix surface. It proved that the dominant wear mechanisms were significantly limited.

Comparing the pure zirconia and zirconia-basing composite materials is visible that composite surface roughness is higher. Anyway, the wear rates for zirconia-basing composites were lower than for pure zirconia material.

Wear properties of both (alumina or zirconia) composite types are distinctly different in spite of wear environment.

Conducted tests established that incorporation of second phase grains into alumina matrix influences wear properties changes in high scale. Changes observed for zirconia based composites are not so spectacular but still significant.

Results of performed wear tests suggest that investigated materials are predicted to work at different environments. The wear resistant parts for work at wet environments seems to be the best area of application for zirconia composites. Alumina based materials show the best properties during dry abrasion.

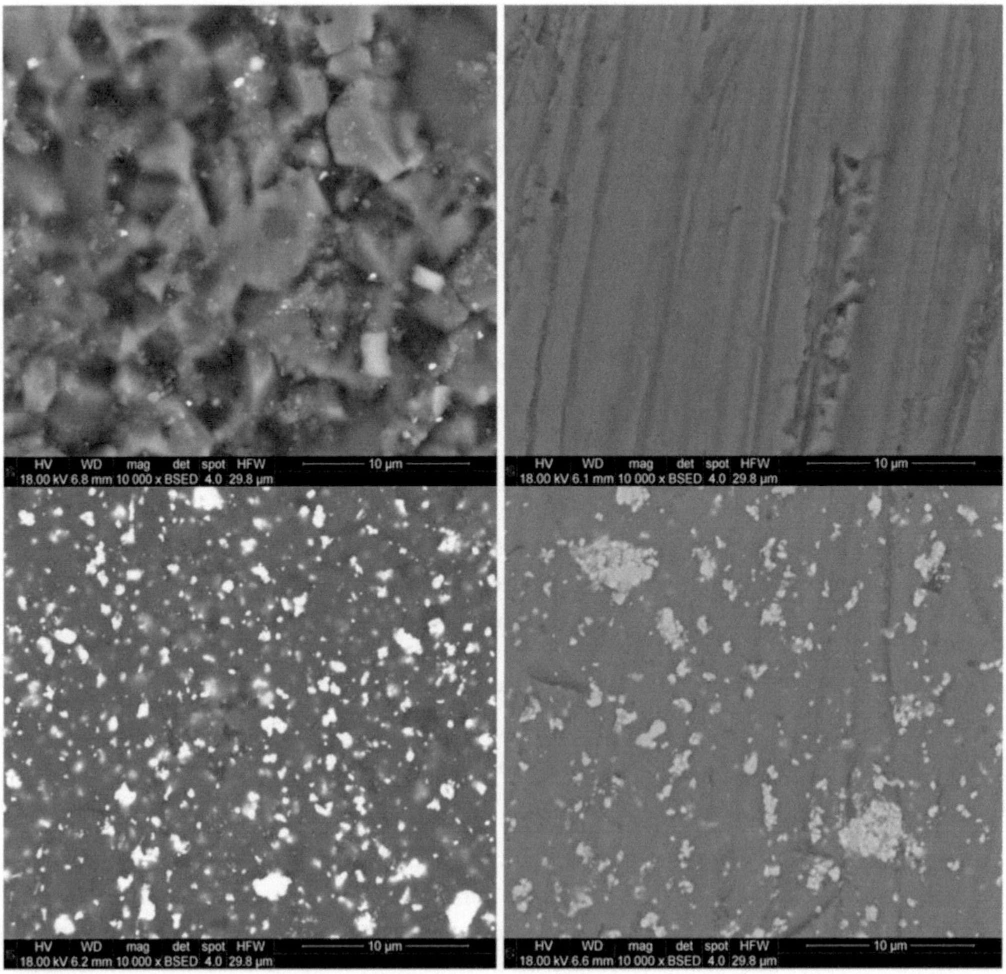

Figure 17. A typical microstructures of worn surfaces after Dry Sand Test; alumina (left top), zirconia s. s. (right top), alumina/10vol.% WC (left bottom), zirconia/10vol.% WC (right bottom).

Material	Miller Test	Dry Sand Test
	R_a [μm]	R_a [μm]
Alumina	0,66 ±0,11	1,02 ±0,05
Alumina/10vol.% WC	0,28 ±0,05	0,50 ±0,02
Zirconia s.s.	0,61 ±0,03	0,71 ±0,04
Zirconia/10vol.% WC	1,00 ±0,14	1,12 ±0,09

± denotes of the standard deviation of 5 measurements

Table 5. Profilographic parameter R_a of material surfaces worn during wear tests.

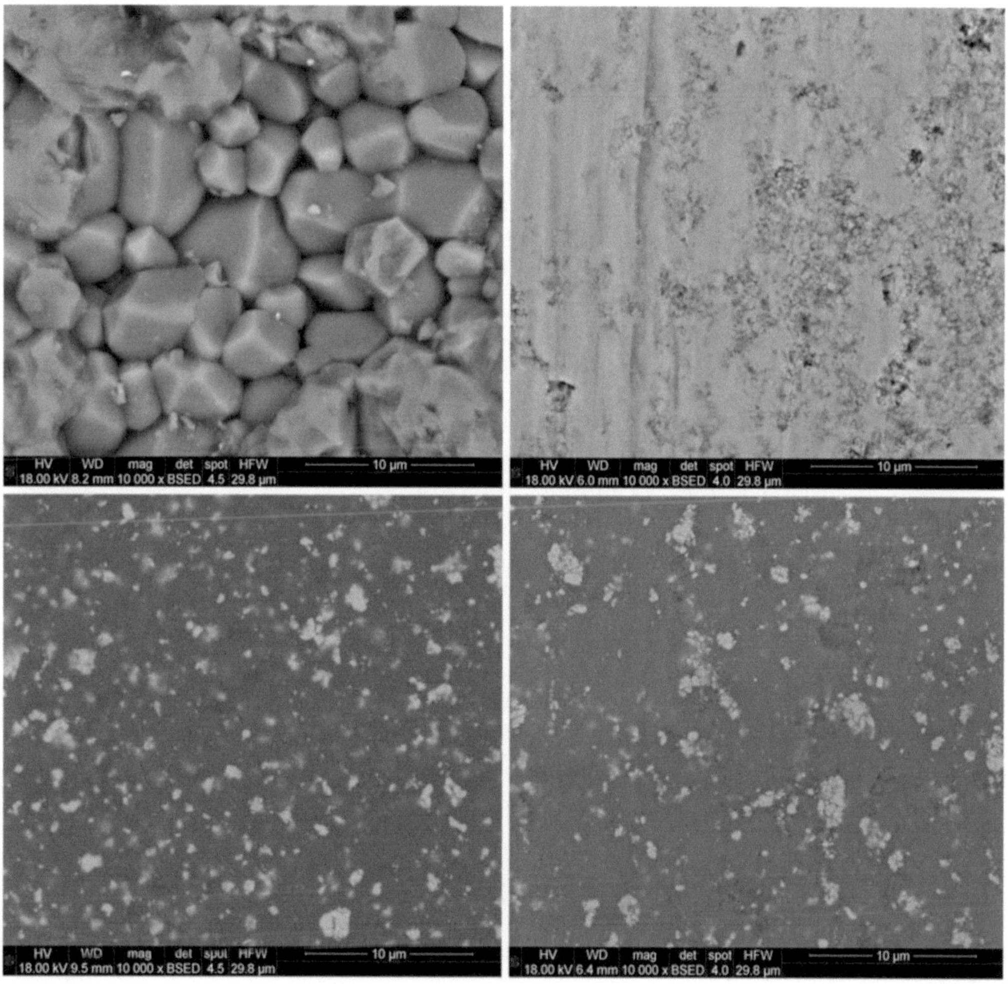

Figure 18. A typical microstructures of worn surfaces after Miller Test; alumina (left top), zirconia s. s. (right top), alumina/10vol.% WC (left bottom), zirconia/10vol.% WC (right bottom).

8. Summary

Selected information about properties of composite materials basing on alumina or zirconia matrices containing dispersed tungsten carbide inclusions presented in this chapter indicated that these materials have potential to be widely used as a structural material.

Properties improvement in these composites is not only an effect of introducing an additional toughening mechanisms connected with crack path/inclusion interacting (crack deflection, crack branching, crack bridging), but also relatively strong interphase grain boundaries confirmed by the unique phenomenon of privileged crystallographic correlation of oxide and carbide phases.

Their very good properties are manifest in applications connected with intensive wear risks, especially in the presence of loose, hard particles. Spectacular improvement was also observed in prolonged applications at conditions under stresses much lower than critical at wet or high humid environments.

Author details

Zbigniew Pędzich
AGH – University of Science and Technology, Krakow, Poland

9. References

[1] Ding, Zh., Oberacker, R. and Thümler, F., Microstructure and mechanical properties of yttria stabilized tetragonal zirconia polycrystals (Y-TZP) containing dispersed silicon carbide particles. J. Europ. Ceram. Soc. 1993; 12(5) 377-383.

[2] Nawa M., Yamazaki K., Sekino T., Niihara K., A New Type of Nanocomposite in Tetragonal Zirconia Polycrystal – Molybdenum System. Materials Letters. 1994; 20 299-304.

[3] Poorteman M., Descamps P., Cambier F., Leriche E., Thierry B., Hot Isostatic Pressing of SiC-Platelets/Y-TZP. J. Europ. Ceram. Soc. 1993; 12 103-109.

[4] Claussen N., Weisskopf K.L. and Ruhle M., Tetragonal zirconia polycrystals reinforced with SiC whiskers. J. Am. Ceram. Soc. 1986; 68 288-292.

[5] Lin G.Y., Lei T. C., Wang S. X. and Zhou Y., Microstructure and mechanical properties of SiC whisker reinforced ZrO_2 (2 mol% Y_2O_3) based composites. Ceramics International 1996; 22 199-205.

[6] Wahi R.P. and Ilschner B., Fracture behavior of composites based on Al_2O_3-TiC. Mater. Sci. Eng. 1980; 15 875–885.

[7] Tiegs T.N. and Becher P.F., Sintered Al_2O_3-SiC Composites. Am. Ceram. Soc. Bull. 1987; 66(2) 339–342.

[8] Niihara K., Nakahira A., Sasaki G. and Hirabayashi M., Development of strong Al_2O_3/SiC composites. MRS Znt. Mtg. on Adv. Mater. Vol. 4 1989; 129-134.

[9] Stadlbauer W., Kladnig W. and Gritzner G., Al_2O_3/TiB_2 composite ceramics. J. Mater. Sci. Lett. 1989; 81217–1220.

[10] Borsa C.E., Jiao S., Todd R.I. and Brook R.J., Processing and properties of Al_2O_3/SiC nanocomposites. J. Microscopy. 1994; 177 305-312.

[11] Breval E., Dodds, G. and Pantano C.G., Properties and microstructure of Ni-alumina composite materials prepared by the sol/gel method. Mater. Res. Bull. 1985; 20 1191–1205.

[12] Sekino T., Nakahira A. and Niihara K., Relationship between microstructure and high-temperature mechanical properties for Al_2O_3/W nanocomposites. Trans. Mater. Res. Soc. Jpn. 1994; 16B, 1513–1516.

[13] Vleugeus J. and Van Der Biest O., ZrO_2-TiX Composites. Key Engineering Materials, Vols. 132-136, Trans Tech Publications, Switzerland, 1997; 2064-2067.

[14] Changxia L, Jianhua Z., Xihua Z., Junlong S., Fabrication of $Al_2O_3/TiB_2/AlN/TiN$ and $Al_2O_3/TiC/AlN$ composites. Materials Science and Engineering A. 2003; 99(1-3) 321-324.

[15] Acchara W., Silva Y.B.S, Cairoc C.A., Mechanical properties of hot-pressed ZrO_2 reinforced with (W,Ti)C and Al_2O_3 additions. Materials Science and Engineering A. 2010; 527 480–484.

[16] Pędzich Z., Haberko K., Babiarz J., Faryna M., TZP - chromium oxide and chromium carbide composites. J. Europ. Ceram. Soc. 1998; 18(13) 1939-1943.

[17] Pędzich Z., Haberko K., Piekarczyk J., Faryna M., Lityńska L., Zirconia matrix - tungsten carbide particulate composites manufactured by hot-pressing technique. Materials Letters. 1998 36(7) 70-75.

[18] Garvie R.C., Stabilization of the Tetragonal Structure in Zirconia Microcrystals. J. Phys. Chem. 1978; 82(2) 218-224.

[19] Kontouros P., Petzow G., Defect Chemistry, Phase Stability and Properties of zirconia Polycrystals. In Science and Technology of Zirconia V. Technomic Publishing Co. Inc. Eds.: Badwall S.P.S., Bannister M.J., Hannink R.H.J. 1992.

[20] Zhang Y., Xu G., Yan Z., Liao C. and Yan C. Nanocrystalline rare earth stabilized zirconia: solvothermal synthesis via heterogeneous nucleation-growth mechanism, and electrical properties. J. Mater. Chem. 2002;12 970-977

[21] Barin I., Knacke O., Thermochemical Properties of Inorganic Substances. Springer-Verlag, Berlin. 1973

[22] Haberko K., Pędzich Z., Piekarczyk J., Bućko M.M., Suchanek W., Tetragonal Zirconia Polycrystals Under Reducing Conditions", Proceedings of Third Euro-Ceramics, vol. 1, Processing of Ceramics, Eds. P.Duran, J.F.Fernandez., Faenza Editrice Iberica S. L., Hiszpania, 1993; 967-972.

[23] Niyomwas S., The Effect of Diluents on Synthesis of Alumina-Tungsten Carbide Composites by Self-Propagating High Temperature Synthesis Process, Proceedings of Technology and Innovation for Sustainable Development International Conference (TISD2010), Faculty of Engineering, Khon Kaen University, Thailand, 4-6 March 2010, 1025-1029.

[24] PTC – the Product Development Company, http://www.ptc.com

[25] Pro/Mechanica – computer program user book (documentation)

[26] Pędzich Z., The reliability of particulate composites in the TZP/WC system. J. Europ. Ceram. Soc. 2004; 24(12) 3427-3430.

[27] Jiang D.T., Van der Biest O., Vleugels J., ZrO_2-WC nanocomposites with superior properties, J. Europ. Ceram. Soc. 2007; 27 1247-1251.

[28] Ünal N., Kern F., Övecoglu M.L., Gadow R., Influence of WC particles on the microstructural and mechanical properties of 3 mol% Y_2O_3 stabilized ZrO_2 matrix composites produced by hot pressing, J. Europ. Ceram. Soc. 2011; 31 2267-2275.

[29] Preiss H., Meyer B. and Olschewski C., Preparation of molybdenum and tungsten carbides from solution derived precursors. J. Mater. Sci. 1998; 33(3) 713-722.

[30] Kim J.C, Kim B.K., Synthesis of nanosized tungstencarbide powder by the chemical vapor condensation process. Scripta Materialia. 2004; 50(7) 969-972.

[31] Rahaman M.N., De Jonghe L.C., Effect of rigid Inclusions on Sintering. Proceedings of First International Conference on Ceramic Powder Processing, Orlando FL. 1-4 November 1987.

[32] Scherer G.W., Sintering with Rigid Inclusions. J. Amer. Ceram. Soc. 1987; 70(10) 719-725.

[33] Weiser M.W., De Jonghe L.C., Inclusion size and Sintering of Composite Powders. J. Amer. Ceram. Soc. 1988; 71(3) C125-C127.

[34] Zhao C., Vleugels J., Basu B., Van Der Biest O., High toughness Ce-TZP by sintering in an inert atmosphere, Scripta Materialia. 2000; 43 1015-1020.

[35] Pędzich Z., Faryna M., Fracture and Crystallographic Phase Correlation in Alumina Based Particulate Composites, in Fractography of Advanced Ceramics II - Key Engineering Materials Vol. 290, Eds. Dusza J., Danzer R. and Morrell R. Trans Tech Publications, Switzerland, 2005; 142-148.

[36] Faryna M., Bischoff E., Sztwiertnia K., Crystal orientation mapping applied to the Y-TZP/WC composite. Microchimica Acta. 2002; 135 55-59.

[37] Sztwiertnia K., Faryna M., Sawina G., Misorientation characteristics of interphase boundaries in particulate Al_2O_3-based composites, J. Europ. Ceram. Soc. 2006; 26 2973-2978

[38] Faryna M., TEM and EBSD comparative studies of oxide-carbide composites. Material Chemistry and Physics. 2003; 81 301-304.

[39] Chevalier J., Saadaoui M., Olagnon C., Fantozzi G., Double-torsion testing a 3Y-TZP ceramic. Ceramics International. 1996;22 171-177.

[40] Evans, A.G., A method for evaluating the time-dependent failure characteristics of brittle materials—and its application to polycrystalline alumina. Journal of MaterialsScience. 1972; 7(10) 1137-1146.

[41] Williams D.P. and Evans A.G., A simple method for studying slow crack growth. J. Testing and Evaluation. 1973; 1 264-270.

[42] Chevalier, J., Olagnon, C. and Fantozzi, G., Subcritical crack propagation in 3Y-TZP ceramics: static and cyclic Fatigue. J. Am. Ceram. Soc., 1999, 82(11), 3129–3138.

[43] Ebrahimi, M.E., Chevalier J. and Fantozzi G., Slow crack-growth behavior of alumina ceramics. Journal of Materials Research. 1999; 15 142-147.

[44] Morita Y., Nakata K., Kim Y.H., Sekino T., Niihara K., Ikeuchi K., Wear properties of alumina/zirconia composite ceramics for joint prostheses measured with an end-face apparatus, Biomed Mater Eng. 2004; 14(3) 263-270.

[45] Pędzich Z., The Abrasive Wear of Alumina Matrix Particulate Composites at Different Environments of Work. In Advanced Materials and Processing IV, Eds. Zhang D. , Pickering K., Gabbitas B., Cao P., Langdon A., Torrens R. and Verbeek J. Trans Tech Publications, Switzerland, 2007; 29-30 283-286.

[46] ASTM 65 - 94 Test Method for Measuring Abrasion Using the Dry Sand/Rubber Wheel Apparatus

[47] ASTM G 75 - 95 Test Method for Determination of Slurry Abrasivity (Miller Number) and Slurry Abrasion Response of Materials (SAR Number).

Chapter 5

Machining Characteristics of Direct Laser Deposited Tungsten Carbide

Paweł Twardowski and Szymon Wojciechowski

Additional information is available at the end of the chapter

http://dx.doi.org/10.5772/51439

1. Introduction

Machinability can be defined as the relative susceptibility of the work material to the decohesion phenomenon and chip formation, during cutting and grinding. This feature depends on work and tool's material physic-chemical properties and condition, method of machining, as well as cutting conditions [1]. Therefore, there is no unique and unambiguous meaning to the term machinability. This feature, can be described by many various indicators. Each one of them carries out a wide variety of operations, each with a different criteria of machinability. A material may have good machinability by one criterion, but poor machinability by another [2].

To deal with this complex situation, the approach adopted in this chapter is to divide machinability indicators into two groups, namely: physical and technological indicators. Physical machinability indicators include i.a. temperatures, cutting forces, vibrations and residual stresses generated during machining process, because their value have the direct influence on the ensemble of the remaining machining effects. Technological indicators include mainly machined surface texture and tool's life (relatively tool wear).

The most popular method for producing tungsten carbide components is by powder metallurgy technology. Nonetheless, for individual, small quantity production or product prototyping this method is too costly and time consuming. The alternative to powder metallurgy is Direct Laser Deposition (DLD) technology, which can be used to quickly produce metallic powder prototypes by a layer manufacturing method [3, 4] – Figure 1. The primary objective of DLD technology is the regeneration of machine parts or machine parts manufacturing with the improved surface layer properties, e.g. higher corrosion, erosion and abrasion resistance. Direct Laser Deposition is an extension of the laser cladding process, which enables three dimensional fully-dense prototype building by cladding consecutive layers on top of one another [6]. The DLD technology is increasingly being used in production of functional prototypes, modify or repair components which have excellent hardness, toughness, corrosion and abrasion wear-resistance, e.g. machine parts for the automotive industry – Figure 2. In the near future DLD technology will be used in manufacturing of spare parts in long term space missions [7] or submarines [8].

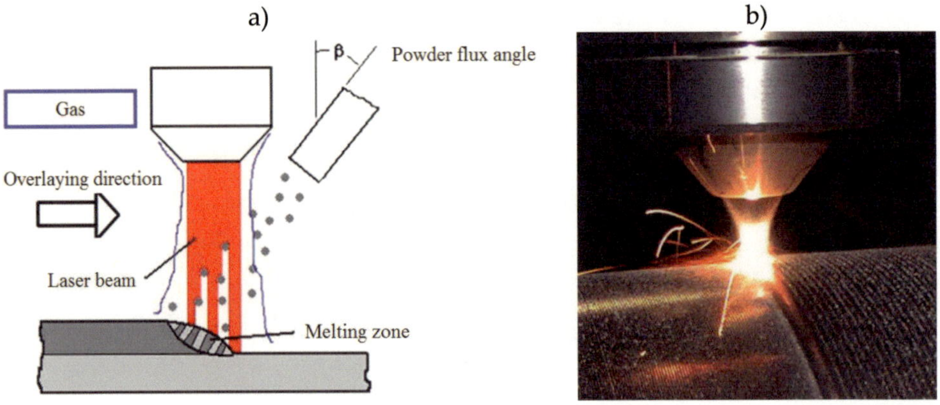

Figure 1. Direct laser deposition technology (DLD): a) the scheme of process, b) the view of process [5]

Figure 2. The application of DLD technology for the crankshafts (a, b) and parts for the automotive industry (c) [5]

Unfortunately, DLD technology has also significant disadvantage. Presently most components produced by DLD technology has an unsatisfactory geometric accuracy as well as surface roughness and requires some post-process machining to finish them to required tolerances [9]. Therefore, the machinability of DLD manufactured materials (e.g. tungsten carbide), require further and extensive studies.

2. Machining of tungsten carbide

Tungsten carbide has excellent physicochemical properties such as, superior strength, high hardness, high fracture toughness, and high abrasion wear-resistance. These properties impinges wide application of tungsten carbide in industry for cutting tools, molds and dies. On the other hand, these unique properties can cause substantial difficulties during machining process, which can result in low machinability. Therefore, machining of tungsten carbide requires the knowledge about the physical effects of the process, as well as appropriate selection of machining method and cutting conditions, enabling desired technological effects. The primary objective of post-process machining of tungsten carbide is to achieve satisfactory geometric and physical properties of its surface texture.

The most popular finishing method of tungsten carbides applied in the tooling industry is grinding with the diamond and CBN (cubic boron nitride) wheels. However, in order to produce optical components made of cemented carbide (e.g. spherical mirrors) the profile quality requires a low surface roughness, a stringent form accuracy on the submicron scale, as well as a low amount of surface damage [10]. Traditional grinding with the diamond wheels can cause machining-induced cracks and damages to the material. To remove these cracks and damage and to obtain a mirror finish, lapping and polishing with fine diamond abrasives are usually employed. Nevertheless, these processes can cause the deterioration of form accuracy and increase the machining cost.

Recently, ultraprecision grinding has been developed that substantially decreases subsurface damage and can precisely control the geometry of the finished surface [11, 12]. This kind of process is conducted on the ultraprecision CNC grinding machines, with three-axes movements, and micro-system to deterministically generate, fine, and pre-polish a plano or spherical surface. Very often these machines have motors with power exceeding 1kW and maximal rotational speeds above 80 000 rpm. The example of ultraprecision set-up is shown in Figure 3a.

Tools applied in the ultraprecision grinding processes are usually selected as metal-bond diamond cup wheels (Figure 3b) with grit sizes between 15÷25 μm. The selected CNC grinding program includes two parts, i.e. stock removal and spark out. During the stock removal step, the grinding speed is selected in the range of 10÷15 m/s (for a small tool diameters it corresponds to rotational speeds up to 40 000 rpm). The vertical feed rates of the tool spindle are usually selected in the range of 0.05÷0.2 mm/min, and the workpiece spindle rotated at 1000 rpm. During the spark out phase, the workpiece is rotated with a 1000 rpm for about 180 rotations.

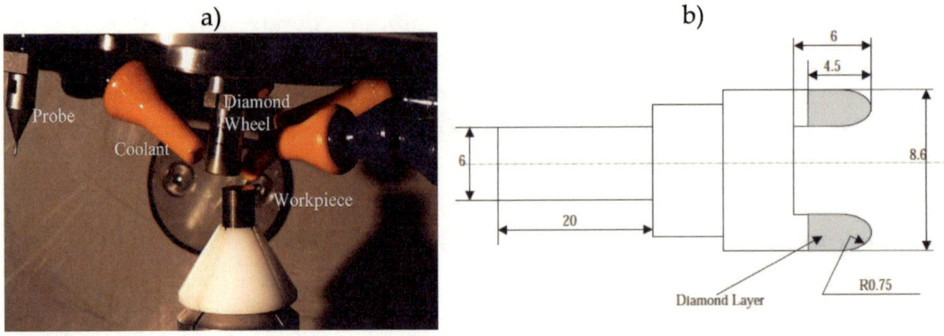

Figure 3. The ultraprecision grinding process of a spherical mirrors: a) set-up [10], b) schematic presentation of the diamond tool [12]

Apart of grinding, recently are seen tendencies to cutting (mainly turning and milling – Figure 4) brittle materials such as, tungsten carbide and reaction-bonded silicon carbide (RB-SiC) by a superhard CBN (cubic boron nitride) and PCD (polycrystalline diamond) cutters in cutting conditions assuring ductile cutting [13, 14]. This technique of cutting can be achieved when depths of cut and feeds (expressed as uncut chip thickness) are extremely low and a quotient of the tool cutting edge inclination angle to uncut chip thickness is greater than unity ($r_n/h>1$). In milling process of tungsten carbide by CBN tools, the transition from ductile to brittle cutting occurs at critical depth of cut $a_{pcr.}$ equal to approximately 4.78 μm. Machining with very low cutting conditions is feasible only on ultraprecision machine tools with high rigidity, which is substantial limitation of this technique.

Figure 4a depicts the schematic diagram of the numerically controlled three-axis ultraprecision lathe used in ductile turning experiments. The lathe has two perpendicular hydrostatic tables along the X- and Z-axis direction, in addition to a B-axis rotary table built into the X-axis table. Both X-axis and Z-axis tables have linear resolutions of 1nm, and the B-axis rotary table has an angular resolution of one ten millionths of a degree. The sample can be rotated with the spindle and moved along the Z-axis direction, while the cutting tool can be moved along the X-axis direction and also rotated around the B-axis.

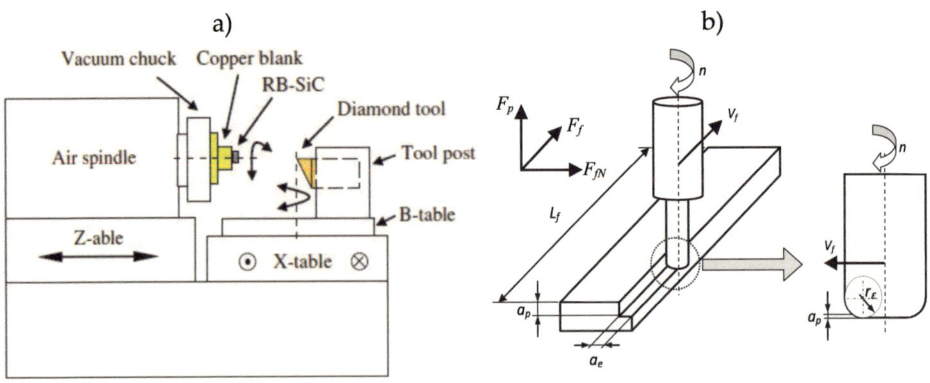

Figure 4. The set-up of cutting process: a) ductile turning of carbides [15], b) face milling of DLD tungsten carbide [16]

Cutters applied in the ductile cutting experiments, are made of diamond (MCD, PCD) or CBN (cubic boron nitride) materials. The example of turning and milling tool applied in carbide's machining process is presented in Figure 5. These tools have usually negative geometry (rake γ angles lower than 0), and a small values of tool cutting edge inclination angle $r_n < 6$ μm, which is needed to initiation of the ductile cutting. To obtain a crack free surface the tool feed rate f and the cutting depth a_p must be very low. Their values are usually selected as: $f \approx 1\div75$ μm/rev and $a_p \approx 2\div10$ μm. Cutting speeds can be selected in the following range: $v_c = 50\div600$ m/min.

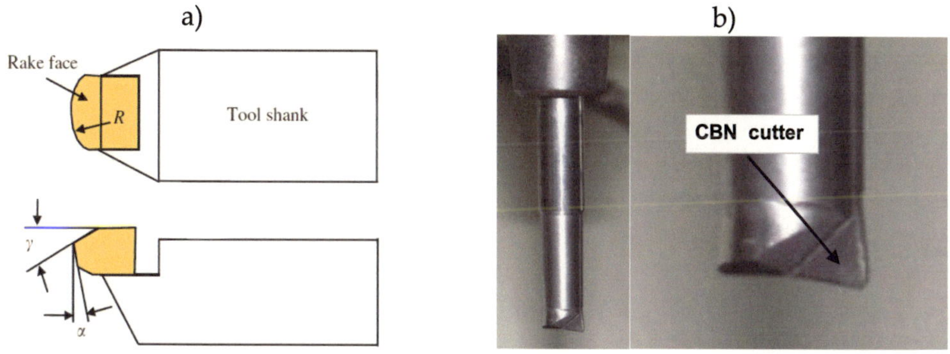

Figure 5. Tools applied in machining of carbides: a) diamond turning tool [15], b) CBN torus end mill [16]

In order to finish plane surface, made of tungsten carbide, obtained using DLD technology, one can apply face milling process (Figure 4b). Surfaces obtained using DLD technology have significantly higher roughness than ones manufactured by powder metallurgy technology. Therefore, cutting parameters during machining of these surfaces can be higher than those applied in machining of powder metallurgy surfaces, and selected as follows: feed per tooth $f_z \approx 25\div100$ μm/tooth, axial depth of cut $a_p = 20$ μm, radial depth of cut $a_e = D/2$ (half of tool's diameter).

3. The analysis of physical machinability indicators

In this chapter the analysis of main physical machinability indicators, such as: cutting forces and vibrations will be presented. The set-up of cutting forces and vibrations measurements during face milling process is presented in Figure 6.

The hook up into bed of a machine piezoelectric force dynamometer was used to measure total cutting forces components [16]. Instantaneous force values were measured in feed force F_f, normal feed force F_{fN} and thrust force F_p directions. Force dynamometer's natural frequency is equal to 1672 Hz. In order to avoid disturbances induced by proximity of forcing frequency to gauge natural frequency, the band – elimination filter was applied. The acceleration of vibrations of tungsten carbide workpiece during milling was measured using piezoelectric accelerometer. These vibrations were measured in the same directions as cutting force components.

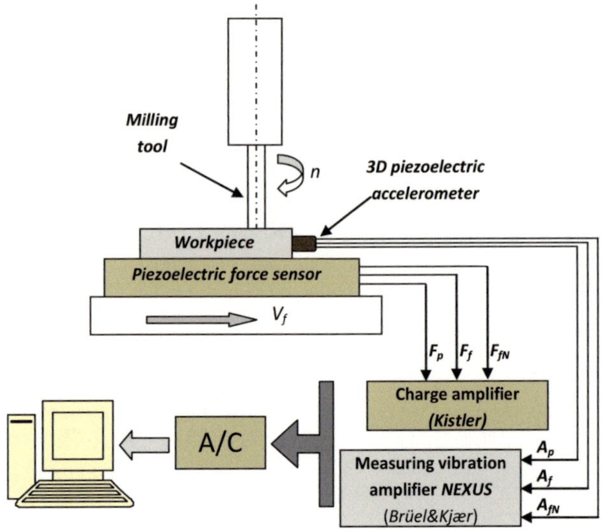

Figure 6. The set-up of force and vibration measurements during face milling of tungsten carbide [16]

Figure 7 depicts the tool wear (*VB_c*) influence on *RMS* values of vibrations A_p and forces F_p.

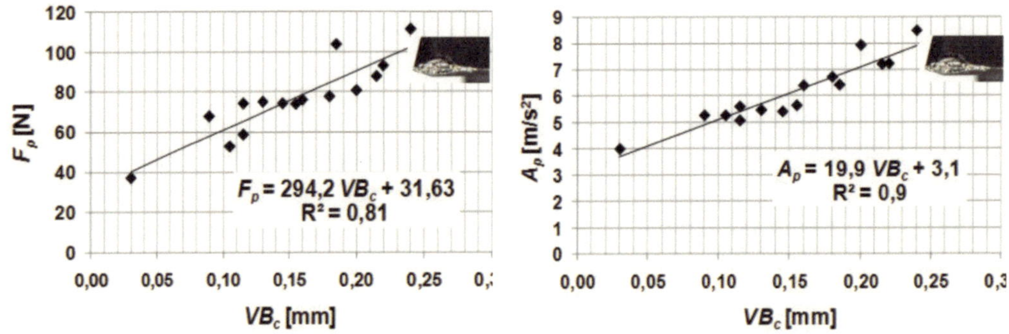

Figure 7. *RMS* values of thrust force F_p and thrust vibrations F_p in function of tool wear VB_c. Cutting conditions: v_c = 68 m/min, v_f = 180 mm/min, a_p = 0.02 mm, a_e = 6 mm

On the base of conducted investigations, clear relation between progressing tool wear and *RMS* values of forces and vibrations in thrust direction (F_p, A_p) can be seen. Above-mentioned relation is expressed by the correlation coefficient R^2>0.8. Tool wear growth induced force F_p and vibration A_p increase, which stays in agreement with investigations [17] related to machining of hardened steel. It was stated that in machining process of tungsten carbide typical abrasion wear, (characterized by VB_c indicator) concentrated mainly on flank face can be found. This phenomenon is probably caused by a friction of hard carbide particles on CBN tool flank face [18]. As a result, progressing abrasion of the tool binder induces the growth of friction force, which in turn is related to force and vibration (F_p, A_p) increase. It is necessary to mention that, in remaining cutting force and vibration directions (F_f, A_f, F_{fN}, A_{fN}) no correlation with tool wear VB_c was found out (correlation coefficient R^2 was lower than 0.1).

In order to analyze forcing frequencies affecting cutting force components during milling of tungsten carbide, the FFT (Fast Fourier Transform) spectra were determined (Figure 8). From the Figure 8 it is resulting, that primary forcing frequency is tooth passing frequency *zfo*. Since number of teeth: *z*=2, the *zfo* frequency overlaps with the second harmonics of spindle speed frequency – 2*fo*. Therefore 2*fo* and *zfo* frequencies are dominant. It means that the dominative factor in F_{fN} and F_p force time courses is milling process kinematics related to the cutting force generated by the each of teeth. Primary harmonic component *zfo* is accompanied by so-called „collateral bands" with the following values: *zfo* + *fo* and *zfo* – *fo*. They appearance is related to the occurrence of radial run out phenomenon. From the Figure 8 it can be also seen that frequency spectra of F_{fN} and F_p force components consist of spindle speed frequency polyharmonics. Similar dependencies were observed for majority of investigated cutting force components frequency spectra.

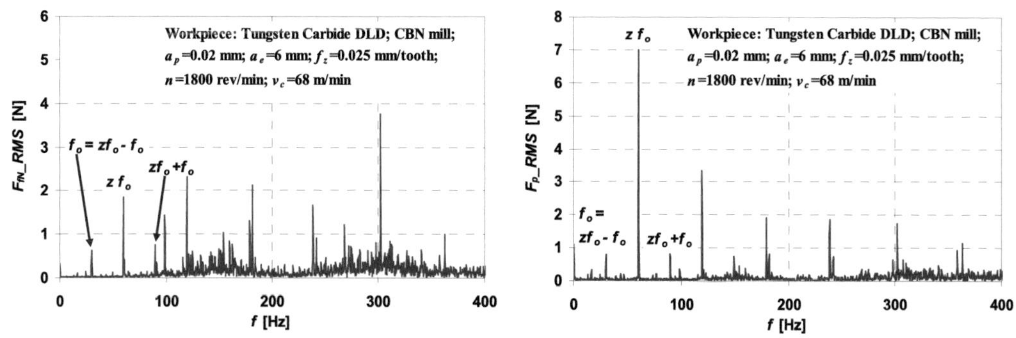

Figure 8. Frequency spectra of F_{fN} and F_p force components

Figure 9 compares cutting forces in function of feed per tooth obtained during milling of tungsten carbide and hardened X153CrMoV12 steel (with 60 HRC hardness). It was observed that both in milling of tungsten carbide and hardened steel, cutting forces (F_f, F_{fN}, F_p) are increasing monotonically with feed per tooth f_z growth, what is typical dependency occurring in metal cutting processes. In tungsten carbide milling process, the highest force values appeared in thrust direction (F_p), independently of feed per tooth f_z value. As it was mentioned before, this phenomenon is probably caused by a friction of hard carbide particles on CBN tool flank face, which affects tool wear increase, and hence friction and thrust force F_p growth (it is worth indicating that after second experimental trial tool wear was $VB_c \approx 0.07$ mm). In case of milling of hardened steel, thrust force F_p has lower values than those obtained during milling of tungsten carbide. This is attributed to the significantly lower hardness of hardened steel (in comparison to tungsten carbide), which reduces tool wear intensity (tool wear after second trial: $VB_c \approx 0$), and thus values of thrust force F_p. The influence of work material's hardness on the cutting force values during machining is described in details in [19].

Figure 10 depicts *RMS* values of vibrations in function of feed per tooth f_z during milling of tungsten carbide.

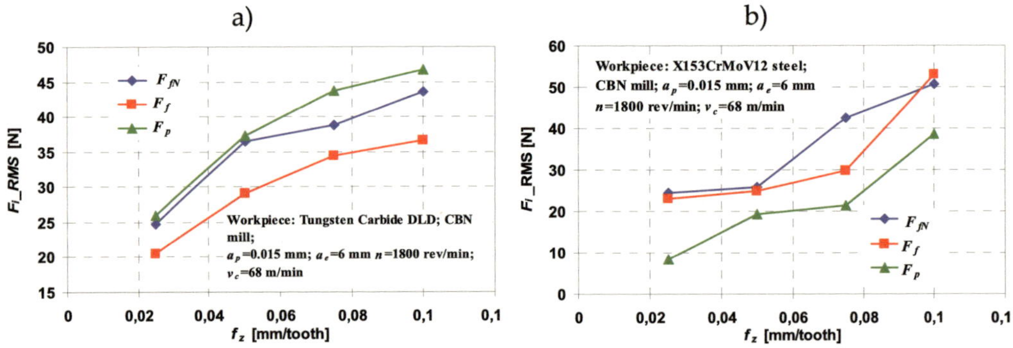

Figure 9. *RMS* values of force components in function of feed per tooth f_z in milling of: a) tungsten carbide, b) hardened steel

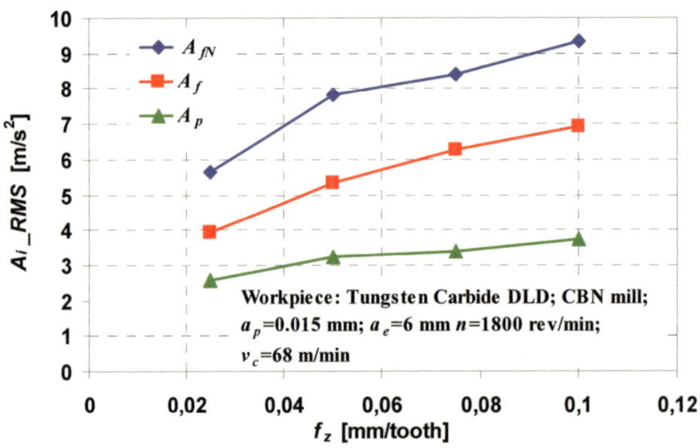

Figure 10. *RMS* values of vibrations (A_f, A_{fN}, A_p) in function of feed per tooth f_z in milling of tungsten carbide

From the Figure 10 it can be seen, that feed per tooth f_z growth induces monotonic increase of vibrations in all measured directions (A_f, A_{fN}, A_p). It was observed that independently of feed per tooth f_z value, vibrations in the thrust direction A_p have the smallest values, in comparison to the other measured directions. The reason of this phenomenon is probably connected with the highest tool stiffness in the thrust direction (parallel to rotational axle). The highest acceleration of vibration values (independently of feed per tooth f_z value) occurred in the feed normal direction A_{fN}. According to research: [20, 21], it could be caused by the direct contact of cutter radius and tool flank face with the machined surface, which is the major source of forcing vibrations, and also the smallest damping ratio in the feed normal direction compared to the other two axes.

In order to estimation of cutting forces in the broad range of cutting conditions, cutting force models can be applied. Majority of models assume that cutting force is proportional to sectional area of cut and the specific cutting pressures. Figure 11 depicts, empirically determined course of the specific cutting pressure in function of mean uncut chip thickness k_i=f(h), in milling of tungsten carbide, while table 1 specific cutting pressure (k_c, k_{cN}, k_p) regression equations.

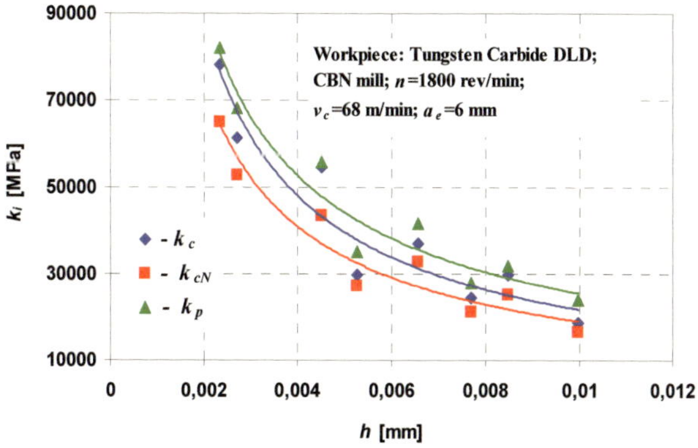

Figure 11. Specific cutting pressure k_i in function of mean uncut chip thickness h

From the Figure 11 it can be seen, that mean uncut chip thickness h growth is accompanied by the specific cutting pressure (k_c, k_{cN}, k_p) decrease. This phenomenon stays in agreement with the dependency observed in metal cutting processes. Between experimental specific cutting pressure values and calculated ones (based on regression analysis) some divergences can be seen. These divergences are expressed by the correlation coefficient R^2>0.87 (table 1). Above-mentioned divergences have disadvantageous influence on mechanistic cutting force model accuracy. The reason of their occurrence could be attributed to milling process dynamics (e.g. machine tool stiffness).

Specific cutting pressure component	Specific cutting pressure k_i	
	Regression equation	R^2
Tangential	$k_c = 388.9\ h^{-0.872}$	0.873
Radial	$k_{cN} = 385.4\ h^{-0.845}$	0.901
Thrust	$k_p = 644.1\ h^{-0.798}$	0.915

Table 1. Regression equations of specific cutting pressure components

4. The analysis of technological machinability indicators

Machined surface texture and tool wear are the essential factors determining cutting ability in practical applications. One of the most popular geometrical tool wear indicators is tool wear on the flank face designated by the *VB*. Its value can be measured using stereoscopic microscopes. The method of tool wear measurement is depicted in Figure 12. Machined surface texture can be examined using three and two dimensional (3D, 2D) measurements. 3D measurements can be achieved using stationary profilometer Hommelwerke T8000 (Figure 13). Two dimensional measurements can be made by T500 profilometer (Hommelwerke), equipped with T5E head and Turbo DATAWIN software. The sampling length lr = 0.8 mm, the evaluation length $ln = 5 \cdot lr$ = 4.8 mm, the length of wave cut – off λc (cut – off) = 0.8 mm and ISO 11562(M1) filter are usually applied in the measurements. As a result of 2D measurements the surface profile charts are received. On the basis of surface profile charts the *Ra* and *Rz* parameters can be calculated using appropriate software.

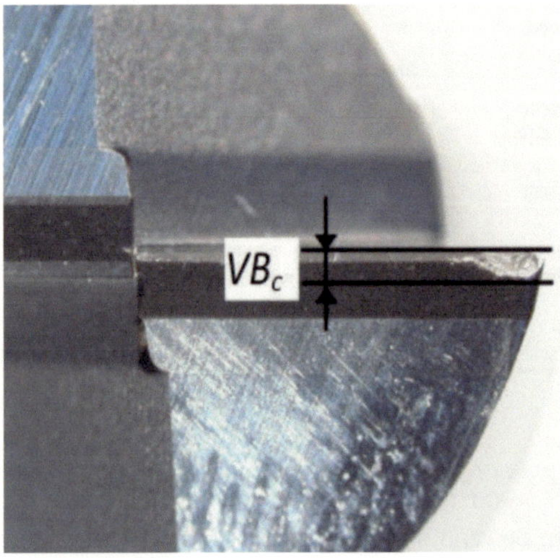

Figure 12. Tool's flank wear measurement

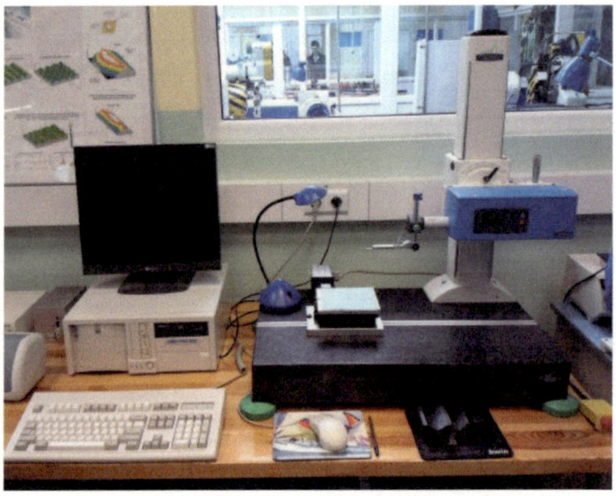

Figure 13. The view of stationary profilometer Hommelwerke T8000

Figure 14 depicts the tool wear progress in function of cutting time during face milling of tungsten carbide (manufactured by DLD technology) with CBN cutters. As it can be seen, tool wear process for each tooth is similar, i.e. there are no significant deviations of VB_c values for respective teeth. Introducing arbitrary dullness criterion $VB_c = 0.2$ mm, it can be seen that twofold cutting speed v_c increase, caused almost eightfold tool life T decrease. On the basis of acquired data the s exponent used in Taylor`s equation ($T = C_T/v_c^s$, where C_T is constant dependent of workpiece properties) can be estimated, but it is necessary to emphasize that determining the s exponent from two experimental values is not very accurate. After consideration of the v_{c1}, v_{c2}, T_1 and T_2 the $s = 2.65$ was obtained. This value is located in the range of the s exponents characteristic for high speed milling of hardened steel, thus the intensity of cutting speed v_c influence on tool life T in tungsten carbide milling is similar to those for hardened steel. Moreover the tool wear concentrates on the flank face of the tool (see Figure 14c). Because of this, the relations between the tool wear and both forces and vibrations in thrust direction were observed (see Figure 7).

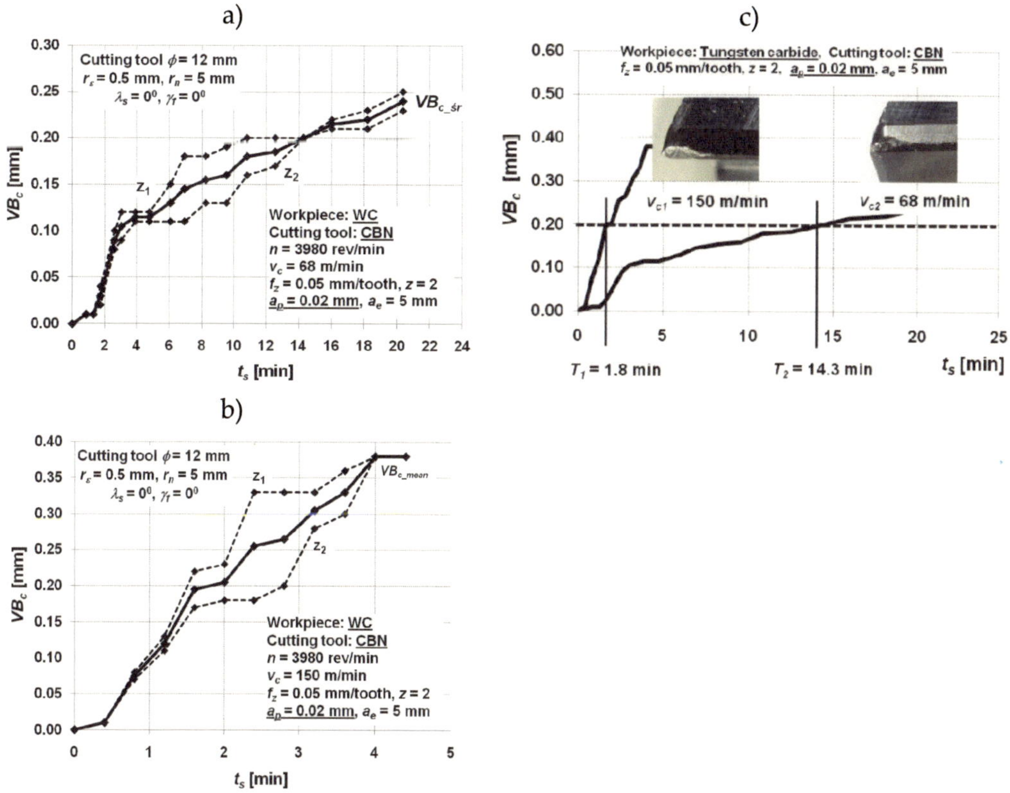

Figure 14. a), b) Tool wear in function of cutting time t_s for two investigated cutting speeds v_c; c) tool wear comparison for exemplary dullness criterion $VB_c = 0.2$ mm (z_1, z_2 – number of tooth, T_1, T_2 – tool life)

Figure 15 compares the surface texture of tungsten carbide sample manufactured by DLD technology before and after milling.

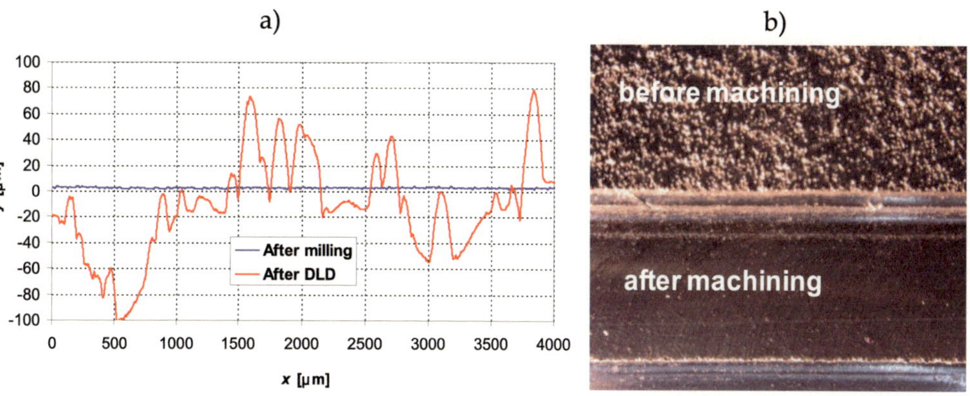

Figure 15. Surface texture of tungsten carbide manufactured by DLD technology, before and after machining: a) 2D surface profile, b) image of surface

It can be seen, that tungsten carbide sample manufactured by DLD technology has an unsatisfactory geometric accuracy and unreasonable surface roughness. Furthermore, from the surface profile and the FFT analysis (Figure 16) it is resulting, that surface texture after DLD process has a random character. The FFT analysis of surface profile consists also of constituent related to the half of the evaluation length (2.4 mm), which means that DLD surface profile is affected by the waviness. Therefore, it needs further finishing process. After milling, machined surface is much smoother and characterized by significantly lower values of surface roughness parameters.

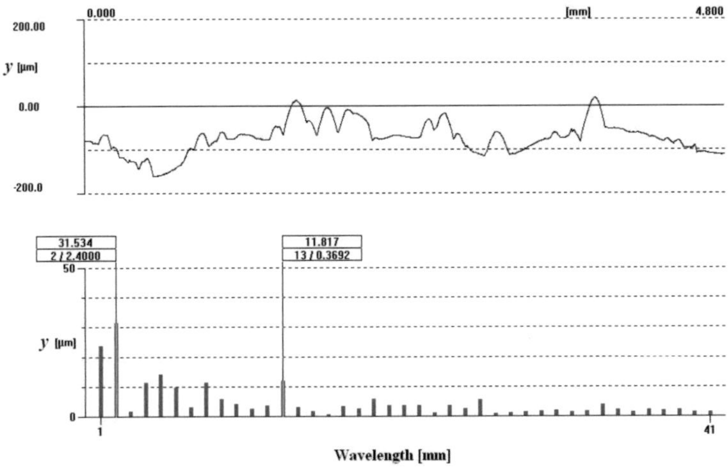

Figure 16. FFT analysis of surface profile after DLD process of tungsten carbide

Figure 17 depicts 3D surface roughness charts and power density spectra (PDS) obtained after milling of tungsten carbide.

It can be seen, that 3D surface topographies after milling (Figure 17) are affected by the cutter's projection into the workpiece. This observation is also confirmed by the power density spectra which represent wavelengths of surface irregularities generated during machining. Surface profiles consist of wavelengths related to the feed per tooth value (f_z = 0.05 mm) which is related to the kinematic-geometric projection of cutter into the workpiece, and feed per revolution value (f = 0.1 mm) which can be induced due to radial run out phenomenon.

Figure 18 depicts examples of profile charts and corresponding to them Ra and Rz parameters for various feed per tooth f_z values. As it can be seen the fourfold feed per tooth f_z increase did not make any significant qualitative and quantitative surface texture changes. It denotes that feed insignificantly influences surface roughness, what is not in full agreement with the results shown in Figures 17a and 17b. For some instances, characteristic kinematic-geometric projection of cutting edge into the workpiece can be seen, however in a wider surface roughness range, there is no typical relation. Figure 19 depicts surface roughness parameters Ra and Rz (for v_c = 68 m/min) in function of feed per tooth f_z.

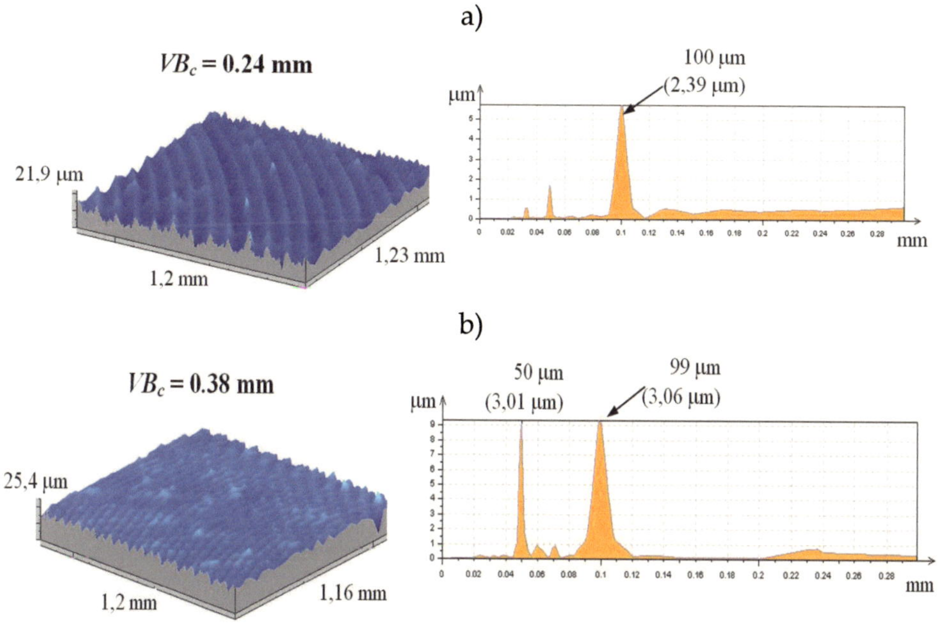

Figure 17. 3D surface roughness chart and corresponding Power Density Function during: a) milling with cutting speed $v_c = 68$ m/min, b) milling with cutting speed $v_c = 150$ m/min

From these charts no influence of feed per tooth f_z on surface roughness is seen, despite for $f_z = 0.1$ mm/tooth. In this case the theoretic value of *Rzt* is comparable to real *Rz* value. It is commonly known that the increase of feed per tooth f_z is accompanied by the increase of surface roughness. Theoretically, the lower feed is fixed, the lower surface roughness is generated. Nevertheless in practice, differences between theoretical and real surface roughness values are increasing with feed decrease. Similar conclusions can be proposed from cumulative *Ra* and *Rz* charts for all a_p and f_z combinations (see Figure 20).

Twofold a_p and fourfold f_z growth caused insignificant *Ra* and *Rz* change. Therefore in the range of conducted research non monotonic increase of surface roughness in function of investigated factors was stated.

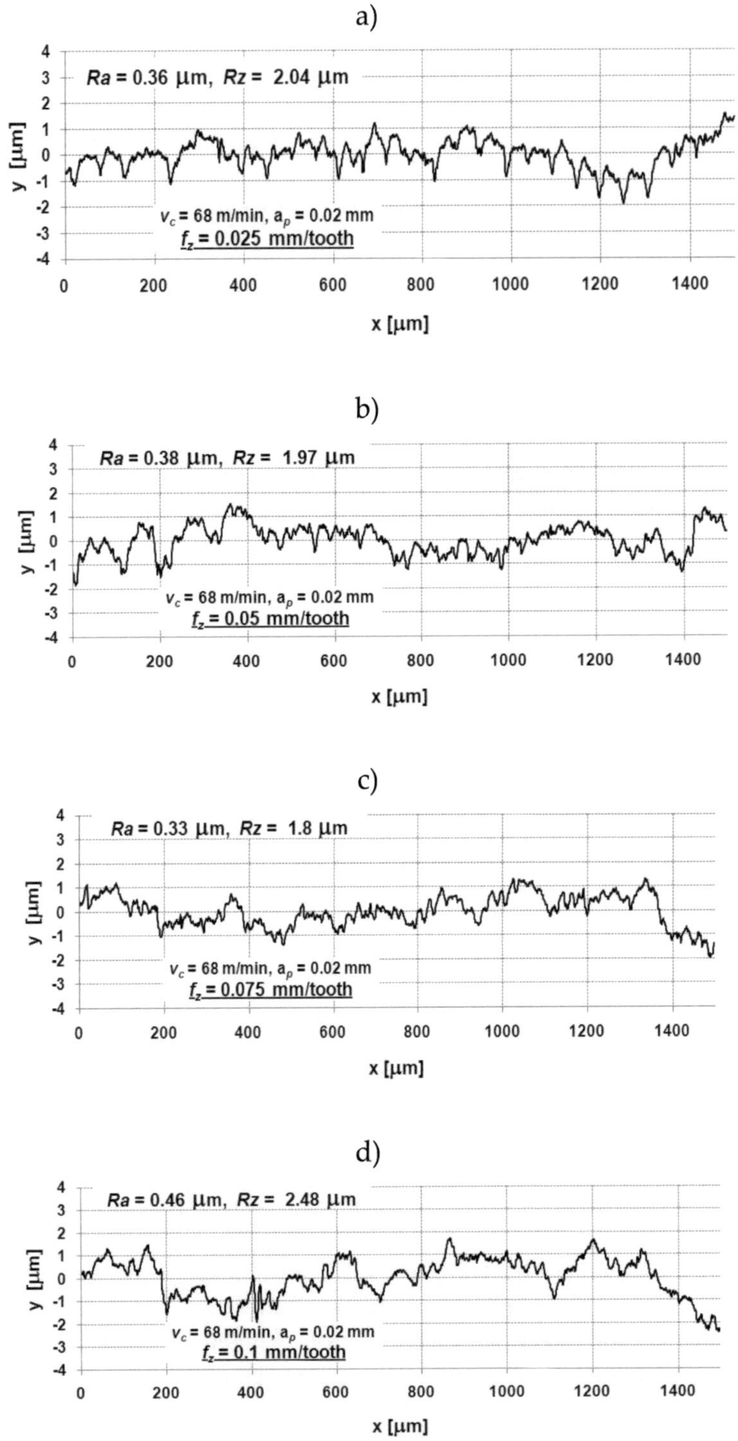

Figure 18. Examples of profile charts for various feed per tooth f_z values: a) f_z = 0.025 mm/tooth, b) f_z = 0.05 mm/tooth, c) f_z = 0.075 mm/tooth, d) f_z = 0.1 mm/tooth

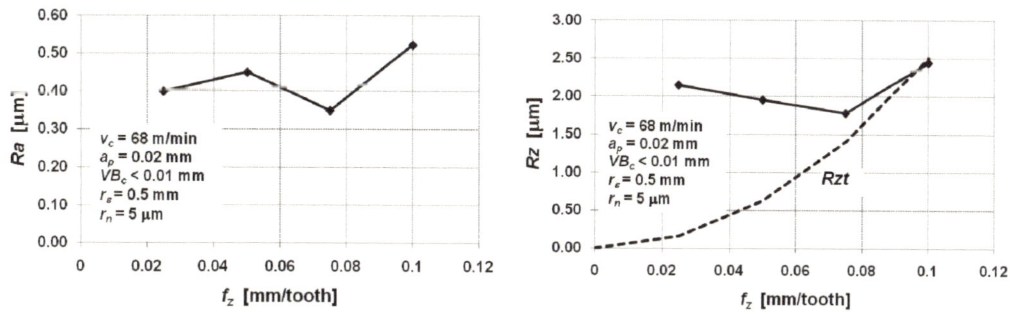

Figure 19. Surface roughness *Ra* and *Rz* in function of feed per tooth f_z

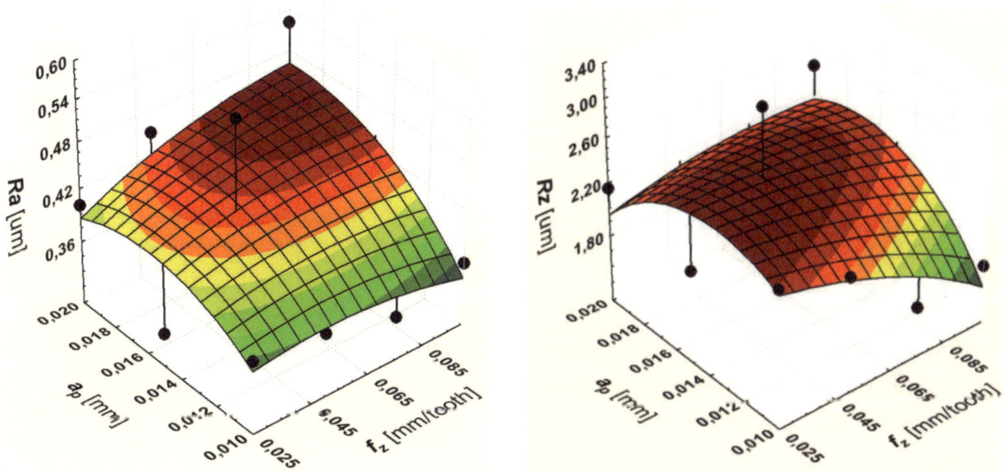

Figure 20. Surface roughness *Ra* and *Rz* in function of feed per tooth f_z and depth of cut a_p

5. Summary and conclusions

The development of modern tool materials such as diamonds (PCD, MCD) and cubic boron nitrides (CBN), as well as ultraprecision and rigid machine tools enables machining of tungsten carbides. These materials have excellent physicochemical properties such as, superior strength, high hardness, high fracture toughness, and high abrasion wear-resistance. On the other hand, these unique properties can cause substantial difficulties during machining process, which can result in low machinability. From the carried out experiments it can be seen, that during machining of tungsten carbides, excessive values of vibrations and intense tool wear growth can occur.

Figure 21 depicts schemes of tungsten carbide products manufacturing processes.

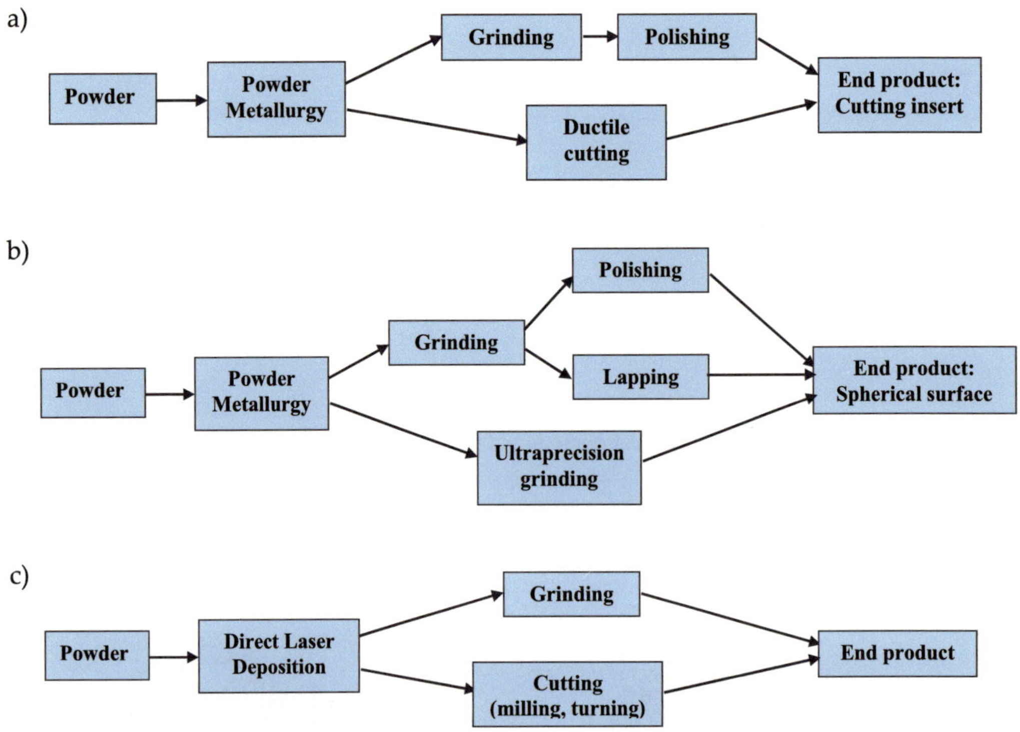

Figure 21. The schemes of manufacturing processes of various products made of tungsten carbide: a) cutting insert, b) spherical surface, c) end product made by DLD technology

The application of ductile cutting to production of cutting inserts (Figure 21a) shortens manufacturing process by the elimination of one partial process (e.g. polishing). However ductile cutting occurs only in the range of extremely low values of depths of cut and feeds. Therefore, this kind of process can be achieved only on very rigid and ultraprecision machine tools, what is substantial limitation of this method. Ultraprecision machine tools can be also applied to grinding of very accurate spherical surfaces. This process also shortens manufacturing process by the elimination of polishing or lapping (Figure 21b). In case of tungsten carbide products obtained by DLD (direct laser deposition) technology (Figure 21c), grinding or cutting (e.g. milling, turning) can be applied as the finishing process. However cutting enables also the shaping of manufactured part, by the possibility of higher cutting conditions application in comparison to grinding. Nevertheless, during cutting of tungsten carbide, intense tool wear growth can occur, and thus this process requires the selection of appropriate cutting conditions.

Deliberations presented in this chapter reveal, that efficient machining process of tungsten carbide parts is feasible, however it requires the knowledge about the physical effects of the process, as well as appropriate selection of machining method and cutting conditions, enabling desired technological effects.

Author details

Paweł Twardowski and Szymon Wojciechowski
Poznan University of Technology, Faculty of Mechanical Engineering, Poznan, Poland

6. References

[1] Schneider G (2002) Cutting Tool Applications. GMRS.
[2] Trent E, Wright P (2000) Metal cutting (4th edition). Butterworth-Heinemann.
[3] Banerjee, R, Collins, P C, Genc A (2003) Fraser, Direct laser deposition of in situ Ti-6Al-4V-TiB composites. Materials Science and Engineering, A358, 343–349.
[4] Fearon E, Watkins K G, (2004) Optimisation of layer height control in direct laser deposition. 23nd International Congress on Applications of Lasers & Electro–Optics (ICALEO 2004), San Francisco, California, Paper No. 1708, Laser Institute of America, Publication No 597, Vol. 97.
[5] http://www.lasercladding.com/
[6] Murphy M, Lee C, Steen W M (1993). Studies in rapid prototyping by laser surface cladding. Proceedings of ICALEO, 882–891.
[7] Taminger K M B, et all (2002). Solid freeform fabrication: An enabling technology for future space missions. Proceedings of International Conference on Metal Powder Deposition for Rapid Manufacturing, San Antonio, 51–60. \
[8] Mazumder J, et al (1999). Direct materials deposition: designed macro– and microstructure. Materials Research Innovations,3, publ. Springer-Verlag, 118–131.
[9] Choi J, Sundaram R (2002). A process planning for 5 – axis laser-aided DMD process. Proceedings of International Conference on Metal Powder Deposition for Rapid Manufacturing, San Antonio, 112–120.
[10] Ling Yin, Spowage A C, Ramesh K, Huang H, Pickering J P, Vancoille E Y J (2004) Influence of microstructure on ultraprecision grinding of cemented carbides. International Journal of Machine Tools & Manufacture 44: 533–543.
[11] Stephenson D J, Vaselovac D, Manley S, Corbett J (2001) Ultra-precision grinding of hard steels. Precision Engineering 25 (4) 336–345.
[12] Yin L, Vancoille E Y J, Lee L C, Huang H, Ramesh K, Liu X D (2004) High-quality grinding of polycrystalline silicon carbide spherical surfaces. Wear 256: 197–207.
[13] Liu K, Li X P (2001) Ductile cutting of tungsten carbide, Journal of Materials Processing Technology, 113: 348-354.
[14] Liu K, Li X P (2001) Modelling of ductile cutting of tungsten carbide, Trans. NAMRI/SME, XXIX 251-258.
[15] Yan J, Zhang Z, Kuriyagawa T (2009) Mechanism for material removal in diamond turning of reaction-bonded silicon carbide, International Journal of Machine Tools & Manufacture, 49: 366-374.
[16] Wojciechowski S, Twardowski P (2010) Cutting forces and vibrations analysis in milling of tungsten carbide with CBN cutters. Proceedings of 4th CIRP International

Conference on High Performance Cutting, 24-26 october, 2010, Nagaragawa Convention Center, Gifu, Japan.

[17] Toenshoff H K, Arendt C, Ben Amor R (2000) Cutting of hardened steel, Annals of the CIRP Vol. 49/2: 546-566.

[18] Liu K, Li X P, Rahman M, Liu X D (2003) CBN tool wear in ductile cutting of tungsten carbide. *Wear*, 255: 1344–1351.

[19] Matsumoto Y, Barash M M, Liu C R, (1987) Cutting mechanisms during machining of hardened steels, Material science and technology, 3, 299 – 305.

[20] Toh C K, (2004) Vibration analysis in high speed rough and finish milling hardened steel, Journal of Sound and Vibration 278: 101–115.

[21] Ning Y, Rahman M, Wong Y S, (2000) Monitoring of chatter in high speed end milling using audio signals method, in: Proceedings of the 33rd International MATADOR Conference, Manchester, England, 421-426.

Chapter 6

Fabrication of Microscale Tungsten Carbide Workpiece by New Centerless Grinding Method

Yufeng Fan

Additional information is available at the end of the chapter

http://dx.doi.org/10.5772/51405

1. Introduction

Recent years have seen the rapid increase in the demand for microscale components smaller than 100μm in diameter, such as micro machine parts, micromachining tools, micro pin gauges, medical catheters, and probes used in scanning tunneling microscope (STM) and semiconductor inspection. To meet this demand, many researchers have actively engaged in the development of new technology for fabricating such devices precisely and efficiently by non-traditional or mechanical machining methods.

Non-traditional machining has employed laser beam lithography and the focused ion beam method. Maruo and Ikuta [1], Yamaguchi et al. [2], and Nakai and Marutani [3] utilized laser beam lithography to fabricate 3D microscale photopolymer components including microscale cylindrical parts. Vasile et al.[4] developed a processing method for the sharpening of STM probes with a focused ion beam. Furthermore, electric discharge machining (EDM) technology is quite effective in micromachining, as seen, for example, in studies on wire EDM of minute electrodes by Heeren et al. [5] and Masuzawa et al. [6,7]. However, these non-traditional methods can only be applied to a limited set of materials, and problems involving machining efficiency and accuracy have not been resolved.

On the other hand, traditional mechanical machining methods, such as cutting and grinding, have also been employed in microscale fabrication. For example, Uehara et al. [8] studied electrolytic in-process dressing (ELID) cylindrical grinding of a micro-shaft, and Okano et al. [9] researched cylindrical grinding of a micro-cylinder. Yamagata and Higuchi [10] developed a four-axis controlled ultra-precision machine and conducted precision turning experiments on a stepped shaft. In these traditional mechanical methods, however, the workpiece is held at its end by a chuck or at both ends by two centers during machining operation. Consequently, it is difficult to perform high-efficiency, high-accuracy machining, especially on microscale cylindrical workpieces with a large aspect ratio because of the low stiffness of the workpiece support mechanism. Fortunately, these problems can be solved if a centerless grinding technique is employed since the workpiece can then be supported along its entire length on a regulating wheel and blade. However, in microscale machining by conventional centerless grinding, an extremely thin blade is required because the blade thickness must be smaller than the workpiece diameter so that the regulating wheel does not interfere with the blade. This necessitates the installation of a costly blade and significantly reduces the stiffness of the workpiece support mechanism. In addition, because of the extremely low weight, the microscale workpiece springs from the blade easily during grinding due to the surface tension of the grinding fluid adhering to the lifting regulating wheel circumference surface. This phenomenon is called "spinning" [11], and causes the grinding operation to fail. However, as will be explained below, these problems would be overcome by employing the ultrasonic-shoe centerless grinding technique developed by the present authors [12–15] in microscale fabrication.

2. Ultrasonic vibration shoe centerless grinding method

Fig. 1(a) illustrates the principle of ultrasonic-shoe centerless grinding where an ultrasonic shoe and a blade are used to support the workpiece and feed it toward the grinding wheel, instead of using a regulating wheel as in conventional centerless grinding (see Fig. 1(b)). In the former case (see Fig. 1(a)), an ultrasonic vibration shoe supports the workpiece and feed it towards the grinding wheel. The rotational speed of the workpiece is controlled by the elliptic motion of the shoe end face. Whereas in the latter case (see Fig. 1(b)), the conventional centerless grinding method includes three basic elements: grinding wheel, regulating wheel and blade. A regulating wheel supports the workpiece together with a blade and to feed the workpiece towards the grinding wheel. The rotational speed of the workpiece is controlled by the rotation of the regulating wheel.

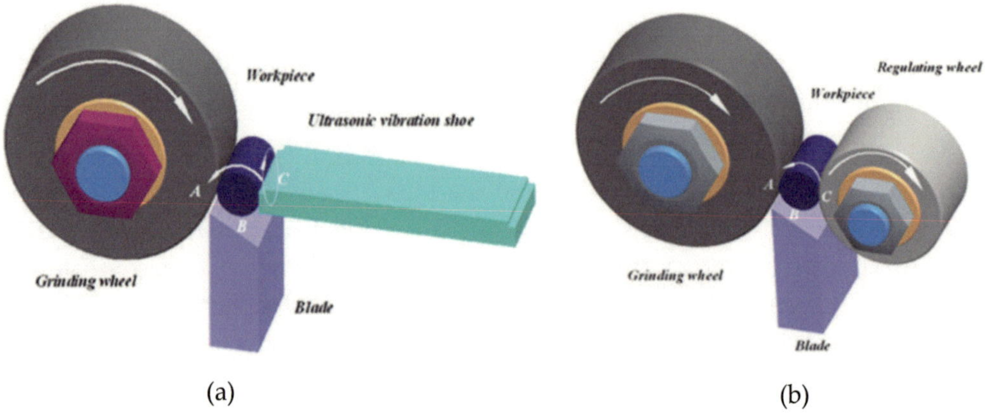

Figure 1. Illustrations of the new centerless grinding with ultrasonic shoe (a) and conventional centerless grinding with regulating wheel (b)

The Fig.2 shows the detail principle of ultrasonic vibration shoe centerless grinding. The workpiece is supported by an ultrasonic elliptic-vibration shoe together with a blade, and it is fed towards the grinding wheel by the shoe. When two alternative current (AC) signals (over 20kHz) with a phase difference of Ψ, generated by a wave function generator, are applied to the PZT after being amplified by means of power amplifiers, the bending and longitudinal ultrasonic vibrations are excited simultaneously. The synthesis of vibration displacements in the two directions creates an elliptic motion on the end face of the metal elastic plate.

Consequently, the rotation of workpiece is controlled by the friction force between the workpiece and the shoe so that the peripheral speed of the workpiece is the same as the bending vibration speed on the shoe end face. The speed varies with the variation of the voltage. In addition, the geometrical arrangements of workpiece such as the shoe tilt angle β, the workpiece center height angle α over the grinding wheel center, and the blade angle ϕ can be adjusted to get the optimum geometrical arrangement in order to achieve the least roundness error.

Based on the processing principle described above (see Fig.2), a grinding apparatus was built as illustrated in Fig.3. The cylindrical workpiece is constrained between the ultrasonic shoe, the blade, and the grinding wheel. The shoe and the blade are fixed on their holders by using bolts. A fine feed mechanism consisting of a linear motion way, a ball screw, and the shoe holder is driven by a stepping motor to give the shoe a fine

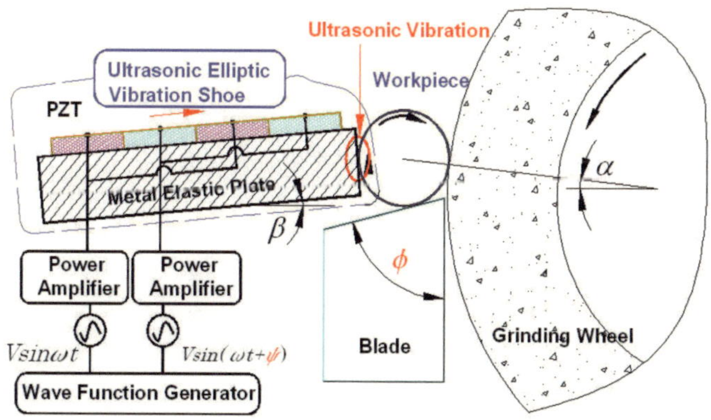

Figure 2. The detail principle of ultrasonic vibration shoe centerless grinding

Motion forward and backward on to the grinding wheel during grinding. The rotational speed of the workpiece is controlled by the elliptic motion of the shoe. Once the clockwise rotating workpiece interferes with the grinding wheel that is rotating counterclockwise at high speed, the workpiece is fed forward and grinding commences. As can be seen in Fig.1 (a), the gap between the lower right edge of the shoe and the top face of the blade should be smaller than the workpiece diameter; otherwise the workpiece would fall through the gap, causing the grinding operation to fail. Therefore, when grinding a microscale workpiece less than100μm in diameter, the vertical position of the shoe must be adjusted carefully so that the gap is sufficiently small. To this end, a fine vertical position adjustment mechanism composed of a vertical motion guide, a ball screw, and a table, on which the fine feed mechanism is held, was constructed in order to adjust the gap by manipulating the ball screw. Moreover, a pre-load is applied to the shoe at its left end face along its longitudinal direction using a coil spring in order to prevent the shoe from breaking due to resonance.

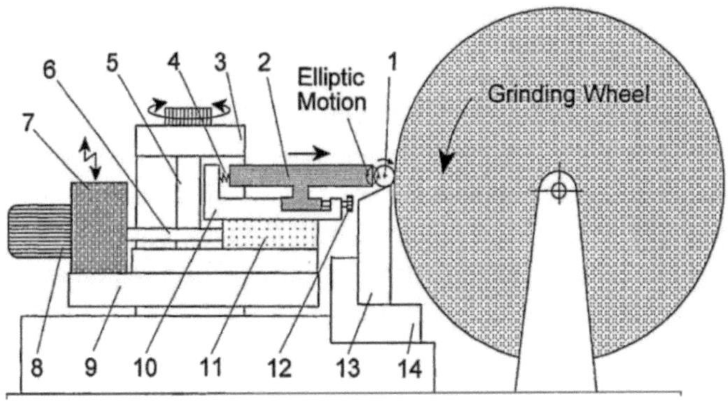

1.Workpiece 2.Ultrasonic shoe 3.Vertical motion guide 4.Pre-load spring 5.Ball screw for adjusting the vertical position of shoe 6.Ball screw for feeding the shoe 7.Gear head 8.Stepping motor 9.Table for holding fine feed mechanism 10.Shoe holder 11.Linear motion way 12.Bolt for fixing the shoe 13.Blade 14.Blade holder

Figure 3. Illustration of the new centerless grinding apparatus

3. Design and construction of an ultrasonic shoe

Fig.4 (a and b) shows the ultrasonic shoe structure and the principle of generation of ultrasonic elliptic motion on the shoe end face, respectively. As shown in Fig.4 (a), the shoe is constructed by bonding a piezoelectric ceramic device (PZT) having four separated electrodes on to a metal elastic body (stainless steel,SUS304). Applying two alternating current voltages (frequency f, amplitude $V_{p\text{-}p}$, and phase difference ψ) generated by amplifying two AC signals from a wave function generator with power amplifiers to the PZT induces simultaneous bending and longitudinal ultrasonic vibrations with amplitudes of several micrometers(see Fig.4(b)). The synthesis of vibration displacements, U_B and U_L, in the two directions creates an elliptic motion on the end face of the metal elastic body. Consequently, the rotation of the workpiece is controlled by the friction force between the workpiece and the shoe (see Fig.1 (a)), and the peripheral speed, V_w, of the workpiece is thus the same as the bending vibration speed, Vs ($=f\times U_B$), on the shoe end face.

Thus, it is essential that the two vibration modes, i.e., bending vibration (B-mode) and longitudinal vibration (L-mode), of the shoe must be induced simultaneously at the same frequency in order to generate an elliptic motion on the shoe end face. The present authors have pointed out in previous studies [12–15] that a combination of an even-ordered B-mode (i.e.,B2, B4,B6, B8, etc.) and an odd-ordered L-mode(i.e.,L1,L3,L5,L7,etc.), where the sole common node for the B-mode and L-mode is located at its central position, should be selected so that the ultrasonic vibrations of the shoe would not be restricted when held at the common node on the shoe holder. In addition, the simpler the vibration mode is, the easier the excitation of the shoe. From this viewpoint, a combination of L1 and B2 modes is desired. However, when the shoe is treated as a plate of length l with a uniform cross-section of width b and thickness t for simplicity, the precondition that the frequency of the r_{th} L-mode must be the same as that of n_{th} B-mode yields the following relationship between l and t [16]:

$$l = \frac{(2n+1)^2 \pi t}{8\sqrt{3}r} \tag{1}$$

Eq. (1) gives the relationship l=5.7t for the L1B2(r=1, n=2) combination, but the relationship l=18.4t for the L1B4(r=1, n=4) combination. This suggests that a thin type shoe, the vibration excitation of which can be more easily compared with others, can be constructed based on the L1B4 combination. Thus, the L1B4 combination was selected as the ultrasonic shoe.

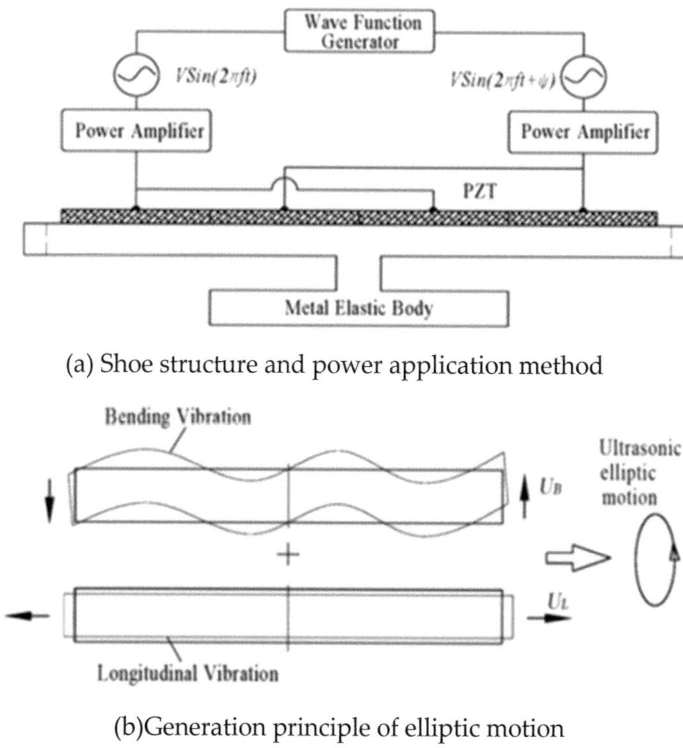

Figure 4. Structure and operating principle of the ultrasonic elliptic vibration shoe

Based on the discussion above, the structure proposed is shown in detail in Fig.5. A T-shaped extrusion is located at the center of the shoe via which the shoe can be fixed on its holder by bolts. Four separate electrodes are distributed on the PZT based on the B4 mode. The dimensions of the shoe are then determined by FEM analysis followed by impedance measurement to be described later.

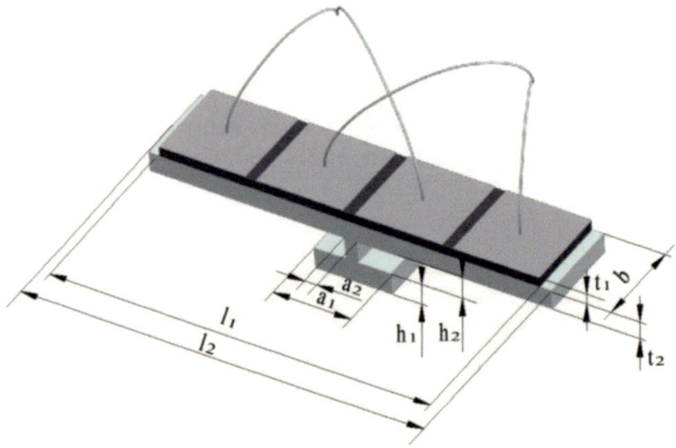

Figure 5. Fig.5 Detailed structure of the ultrasonic shoe

b	t_1	t_2	l_1	l_2	h_1	h_2	a_1	a_2
20	2	4	88.6		3	20	20	5

Table 1. The dimensions of shoe designed (mm)

(a) L1 mode (f_{L1} =23.85 kHz) (b) B4 mode (f_{B4} =23.85 kHz)

Figure 6. L1 and B4 modes obtained by FEM analysis

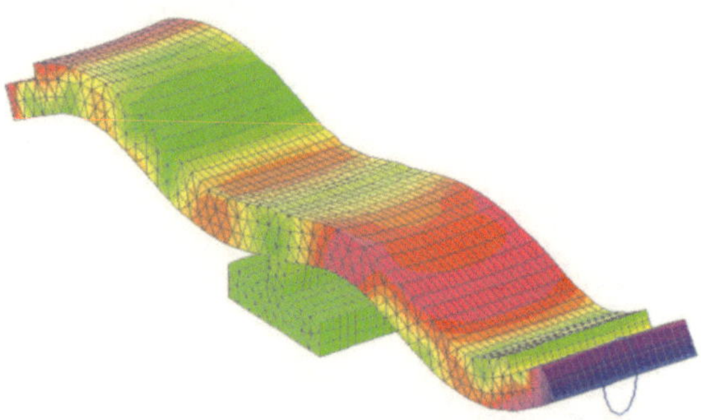

Figure 7. Elliptic motion predicted by FRA (frequency response analysis)

With the exception of the length of the metal elastic body, l_2, all dimensions were determined (see Table1) by taking into consideration the space available for installation of the proposed shoe on the existing centerless grinder. Dimension l_2 was first predicted by finite element method (FEM) analysis under the condition f_{L1} (frequency of L1 mode) = f_{B4} (frequency of B4 mode). Fig.6 (a and b) shows, respectively, the L1 and B4 modes of a shoe (l_2 =96.95mm) obtained by FEM analysis for f_{L1} = f_{B4} =23.85 kHz. In order to confirm the generation of elliptic motion on the shoe end face having the FEM predicted dimension l_2 =96.95mm, a frequency response analysis (FRA) was carried out using piezoelectric device analysis software. Fig.7 shows the FRA results obtained for V_{P-P}=50V, f =23.90 kHz, and ψ =90°. Clearly, an elliptic motion occurs on the end face of the shoe.

As predicted by FEM and FRA above, l_2 must be 96.95mm in order for f_{L1} to equal f_{B4} and for an elliptic motion to be generated on the shoe end face. However, it is foreseen that the actual values of f_{L1} and f_{B4} would not agree with the predicted values due to dimensional errors associated with the metal elastic body and the PZT used. Thus, three shoes with different values of l_2, namely 96.45, 96.95, and 97.45mm, were constructed based on the FEM and FRA results. One of these values was selected after the shoes' actual frequencies f_{L1} and f_{B4} were obtained by measuring their impedance characteristics. Fig.8 shows a photograph of a designed and constructed ultrasonic shoe. The shoe surface was coated with a waterproofing layer in order to protect against the grinding fluid during grinding. Further, the friction coefficient between the shoe and the workpiece should be large enough to prevent the workpiece from slipping on the shoe end face. Thus, a thin rubber(0.5mm in thickness) sheet made of the same materials that used in conventional regulating wheels was prepared and attached to the end face of the shoe.

Figure 8. Photograph of a fabricated shoe

An impedance analyzer was used for investigating the impedance characteristics of the shoes. The results obtained for the shoe having an $l2$ of 96.95mm are shown in Fig.9 (a and b) for the L1 and B4 modes, respectively. Clearly, the impedances for the B4 and L1 modes reach their minima at the frequencies of 24.13 and 24.01kHz, respectively, indicating that the respective resonant frequencies for the L1 and B4 modes are f_{B4} =24.13kHz and f_{L1} =24.01kHz.The impedances for the two modes reach their maxima at 24.20 and 24.22kHz, respectively, meaning the power consumption would be least when the AC voltages applied at these frequencies. This is referred to as the anti-resonance effect [17, 18]. The measured f_{L1} and f_{B4} are plotted against l_2 (Fig.10). It can be seen that f_{L1} comes closest to f_{B4} at l_2 =96.45mm. Thus, l_2 was determined to be 96.45mm.

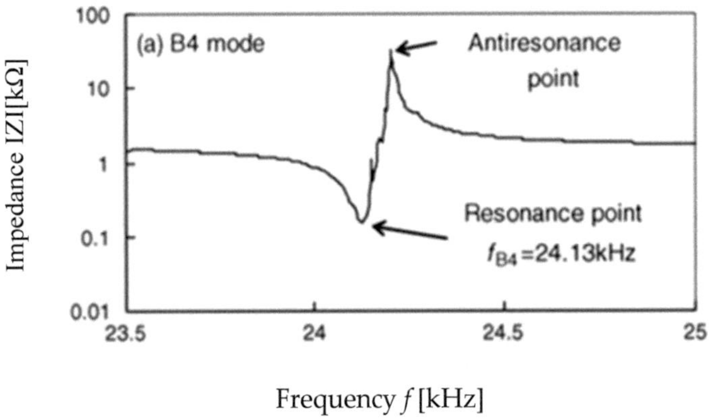

(a)Impedance characteristics of B4 mode

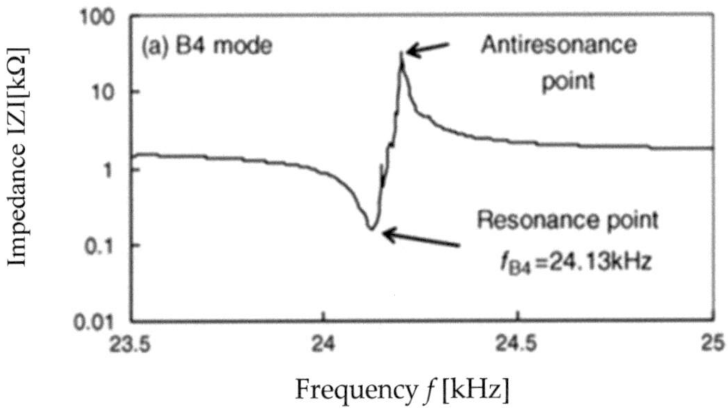

(b)Impedance characteristics of L1 mode

Figure 9. Impedance characteristics of the shoe

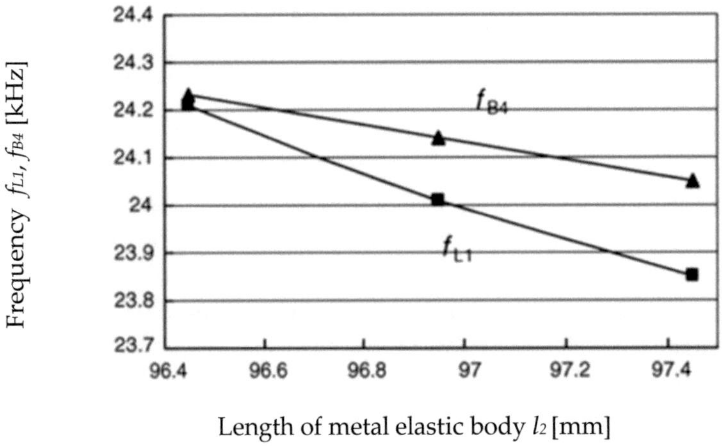

Figure 10. Measured frequencies of two modes

4. Performance of the apparatus constructed

4.1. Method of measuring the ultrasonic elliptic vibration

The elliptic motion of the shoe end face under various applied voltages (amplitudes, frequencies and phase differences) is investigated using a measuring system composed of two laser Doppler vibrometers (Ono Sokki Co., Ltd., LV-1610) equipped with the respective sensor heads, a vector conversion unit (Ono Sokki Co., Ltd.,), and a multi-purpose FFT(Fast Fourier Transform) analyzer (Ono Sokki Co., Ltd., CF-5220), as shown in Fig.11.

The shoe is bolted at its center (the common node for L1 and B4 mode) on the holder in order not to restrict the ultrasonic vibration. A preload is then applied to the shoe using a coil spring in order to prevent the PZT from breaking due to resonance. Two AC signals generated by a wave function generator (NF Corporation, WF1994) are applied to the PZT after being amplified by two power amplifiers (NF Corporation, 4010). During measurement, the two laser beams from the respective heads are focused at the same point near the shoe end face. The signals from the laser Doppler vibrometers are then input to the vector conversion unit for synthesis and are recorder with a digital oscilloscope (Iwatsu Co., Ltd., LT364L). The AC signal is changed by various voltages, phase differences and frequencies. From the digital oscilloscope, the trace of ultrasonic vibration will be obtained based on the different input parameters, and the relationship between the input parameters and the vibration will be clarified.

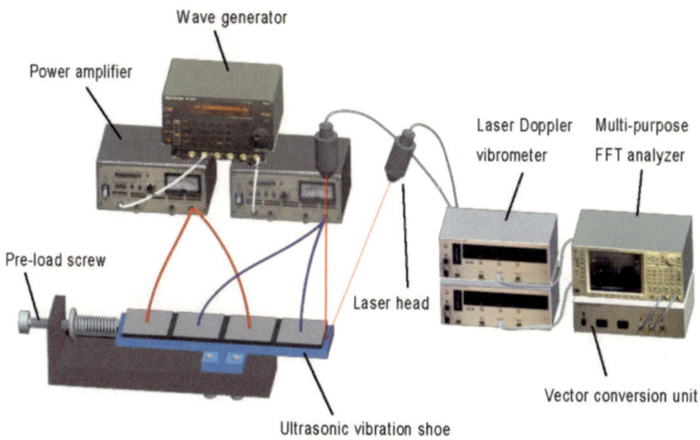

Figure 11. Method of measuring the ultrasonic elliptic vibration

The shoe is bolted at its center (the common node for L1 and B4 mode) on the holder in order not to restrict the ultrasonic vibration. A preload is then applied to the shoe using a coil spring in order to prevent the PZT from breaking due to resonance.

Two AC signals generated by a wave function generator (NF Corporation, WF1994) are applied to the PZT after being amplified by two power amplifiers (NF Corporation, 4010). During measurement, the two laser beams from the respective heads are focused at the same point near the shoe end face. The signals from the laser Doppler vibrometers are then input to the vector conversion unit for synthesis and are recorder with a digital oscilloscope (Iwatsu Co., Ltd., LT364L). The AC signal is changed by various voltages, phase differences and frequencies. From the digital oscilloscope, the trace of ultrasonic vibration will be obtained based on the different input parameters, and the relationship between the input parameters and the vibration will be clarified.

Fig.12 shows the measured results of the point on the end face with various parameters.

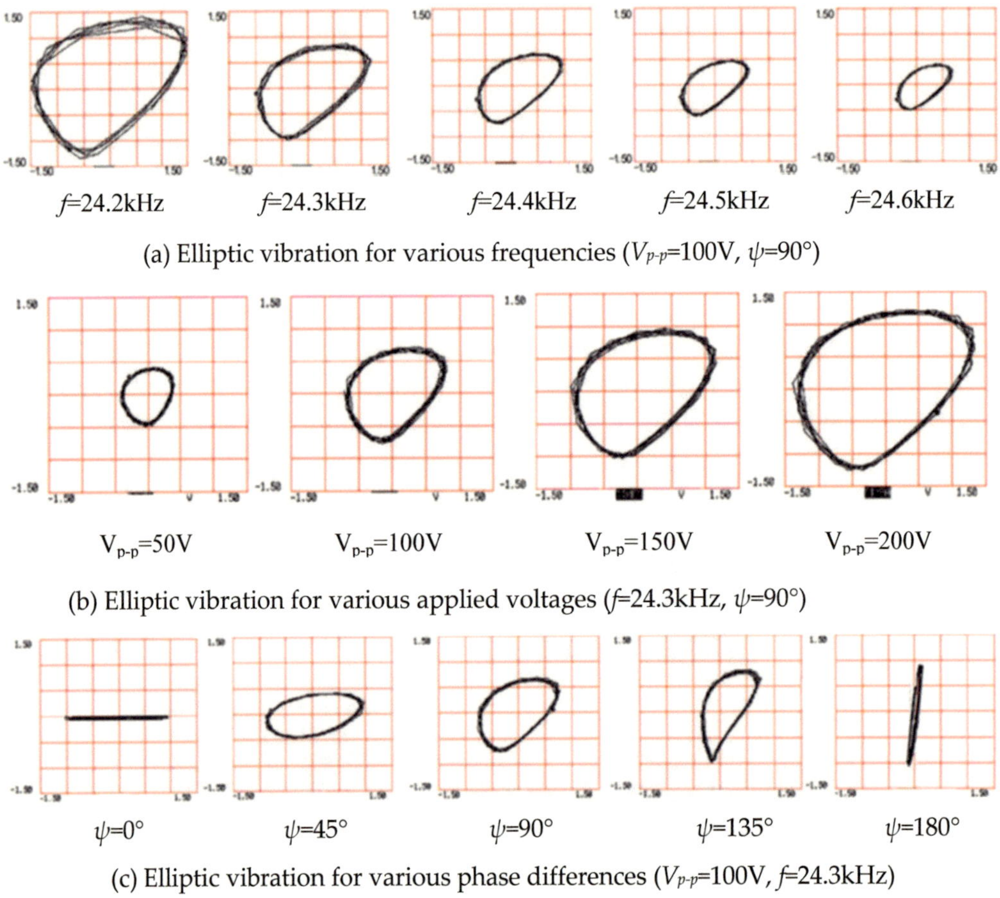

Figure 12. Measured results of the point on the end face with various parameters

4.2. Rotational motion control tests of the workpiece

In the ultrasonic vibration shoe centerless grinding method, it is crucial to precisely control the workpiece rotational speed by the elliptic motion of the end face of shoe in order to achieve high-precision grinding. Therefore, a evaluating involving the rotational control of a cylindrical workpiece using the produced ultrasonic vibration shoe was conducted on an apparatus specially built in house, as shown in Fig.13.

In the apparatus, a wheel mounted on a spindle is driven rotationally by a motor and plays the role of the grinding wheel. The ultrasonic vibration shoe is bolted on its holder and then held on a small 2-axis dynamometer (Kistler Co., Ltd., 9876) installed on a linear motion guide. A thin rubber sheet (0.5mm in thickness) of the same material as that of a conventional regulating wheel (A120R) was made and attached to the shoe end face so that the friction coefficient between the shoe and the workpiece is large enough to prevent the workpiece from slipping on the shoe end face. The workpiece is fed toward the wheel by the shoe, which is carried forward by manipulating the shoe feed bolt. The normal contact force and the friction force between the rotating workpiece and wheel correspond to the normal and tangential grinding forces, respectively. In the test, the dynamometer was used to set up the force, and the same wave function generator and power amplifiers as used in the elliptic motion measurement were employed to apply the AC voltage to the PZT. The workpiece rotational speed is obtained by recording the motion of the rotating workpiece end face, on which a circular mark was created, suing a digital video camera. The video images are then stored in a computer for analysis using animated image processing software (Deigimo Co., Ltd., Swallow2001 DV). Pin-shaped rods (SK4) of 5mm in diameter and 15mm in length were used as the workpieces. In addition, $V_{p\text{-}p}$ was set in the range of 20-200V while the voltage frequency and the phase difference were fixed at f=24.3kHz and ψ=90°, respectively.

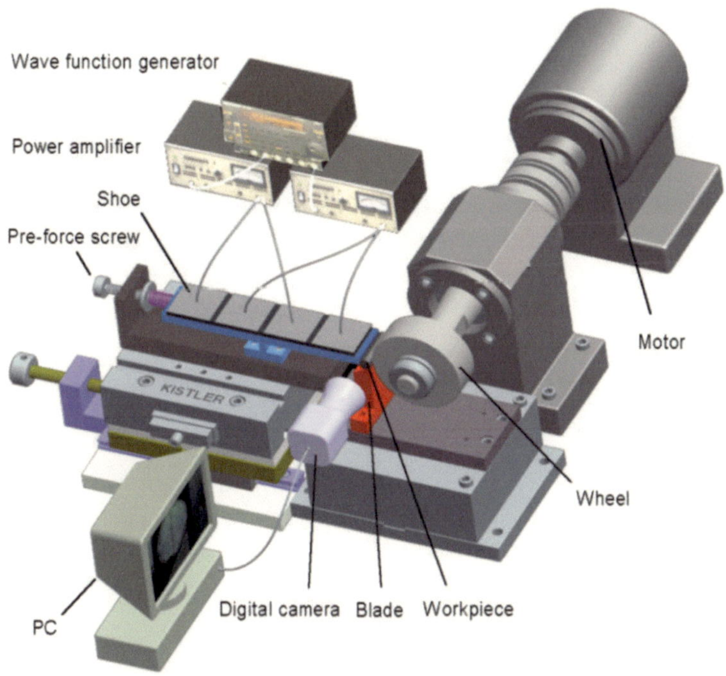

Figure 13. Evaluation apparatus for the shoe

Fig.14 shows a series of video images of the workpiece end face taken every 0.033s with a camera capable of taking 30 pictures per second. The workpiece rotational speed n_w can thus be calculated as follows:

$$n_w = \frac{\sum_{i=1}^{N} n_{wi}}{N} \tag{2}$$

where $n_{wi} = (\beta_{i+1} - \beta_i)/(t_{i+1} - t_i)$, i =1,2, ..., N.

Fig.15 shows the relationship obtained between n_w and V_{p-p}. Clearly, n_w increases linearly with V_{p-p}. This is in close agreement with the prediction described above, and indicates that the workpiece rotation speed can be precisely controlled by the elliptic motion of the shoe.

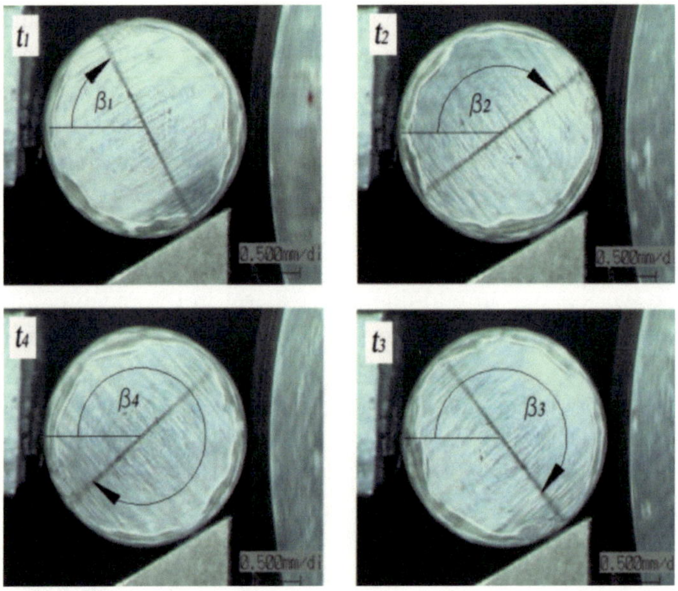

Figure 14. Video images of the rotating workpiece

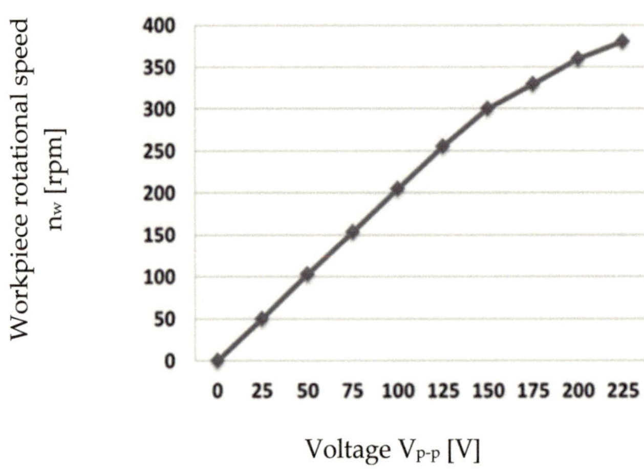

Figure 15. Relationship between the workpiece rotational speed and the applied voltage

5. Fabrication of micro-part of tungsten carbide

5.1. The modification of experimental grinder

In order to confirm the validity of the proposed new method, fabrication of micro-part of tungsten carbide will be carried out. The grinder is modified by the conventional grinder of μ micron grinder MIC-150, the product of μ micron Corp. The regulating wheel unit will be uninstalled and a fine feed unit, which is composed of a fine feed table and stepping motor, will be installed. In the experimental grinder, for the finish grinding of micro parts with sizes of less than 1mm in diameter, the depth of cut must be less than 1μm in order to make the grinding force small. The fine feed and fine adjustment unit is shown in Fig.16 (a and b). The shoe can be fine adjusted in Z direction by handing the fine adjustment screw. The adjustment component can be locked when the height of shoe is adjusted to an appreciable position by operating the lock handle on the back of the unit, as shown in Fig.16 (b). The fine feed and fine adjustment unit can be rotated in XY by surrounding the rotating pin, and then fixed the unit by locking other three fixed screw bolts. A fine feed unit composed of a shoe holder, a linear guide, a ball screw and a stepping motor has been designed and produced that carries the shoe toward the grinding wheel at a feed rate of less than 1μm. A pre-load is then applied to the shoe at its left end face in its longitudinal direction using a coil spring in order to prevent the shoe from breaking due to resonance and is fixed by the screw at its right face of shoe foot.

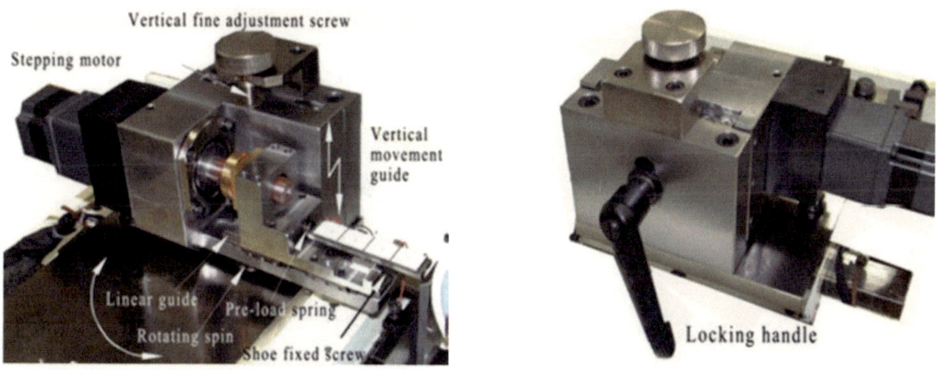

(a) (b)

Figure 16. Fine feed and fine adjustment unit

5.2 Grinding experiments

The grinder installed fine feed and adjustment mechanisms was used to grinding micro-scale cylindrical workpiece, its aim is to verify the feasibility of micro-scale fabrication by ultrasonic-shoe centerless grinding technique, and to confirm the performance of the constructed experimental apparatus in actual grinding operations. The tungsten carbide steel cylindrical workpiece used in grinding is shown in Fig.17, 0.6mm in diameter and 15mm in length. The photo of grinder is shown in Fig.18 and Fig.19 shows a main portion of the experimental setup. The experimental conditions are listed as in Table 2.

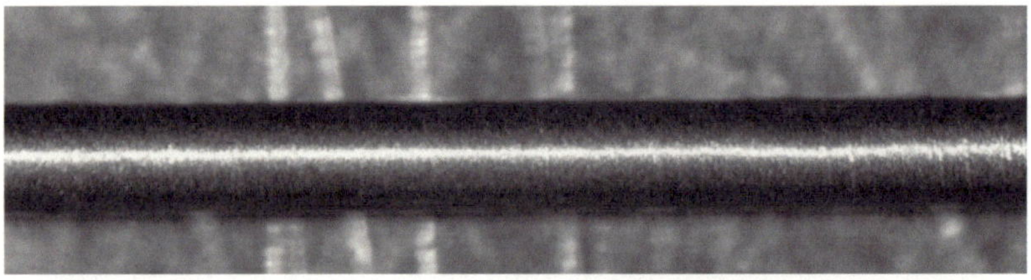

Figure 17. Original tungsten carbide steel cylindrical workpiece (D0.6mm×L15mm)

Figure 18. Photo of grinder installed a fine feed and adjustment mechanisms

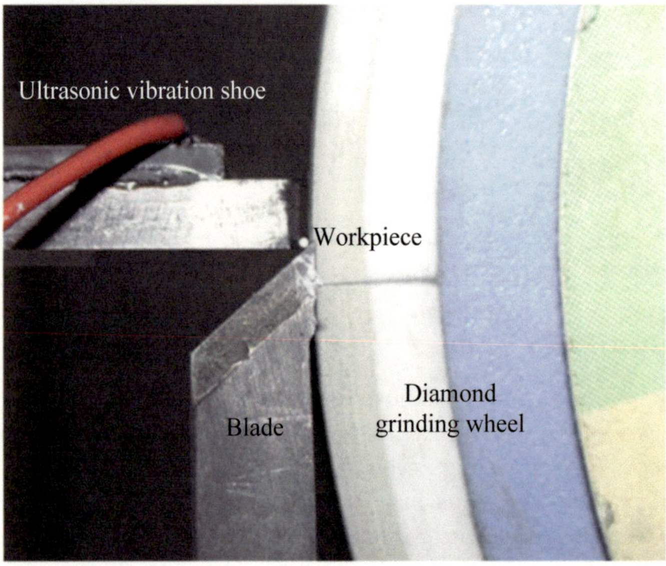

Figure 19. Experimental setup for the grinding test

The grinding test was performed as follows: first, the gap between the right-down edge of the shoe and the top face of the blade was set up carefully using the fine vertical position adjustment mechanism so that the gap is smaller than the final diameter of the workpiece after grinding. Next, the grinding wheel was moved toward the blade and stopped soon after they have interfered with each other. Subsequently, the shoe was carried forward to feed the workpiece toward the grinding wheel at a feed rate of V_f=0.1mm/min and ground under the grinding conditions listed in Table 2. The grinding operation was finished once the given stroke removal and spark out had been completed and the shoe had been then retracted from the grinding wheel.

Grinding wheel	SD2000150×20×76.2,1A1		
Workpiece	Tungsten carbide steel ϕ0.6×L15		
Coolant	Solution type		
Grinding parameters	Input Voltage	Amplitude	$V_{p\text{-}p}$=100V
	Input Voltage	Frequency	f=24.3kHz
	Input Voltage	Phase difference	90°
	Grinding wheel speed		V_g=30m/s
	Shoe feed rate		V_f=0.1mm/min
	Stock removal		0.4-0.55mm
	Spark-out time		3 sec.
Geometrical conditions	Center height angle		γ=7°
	Blade angle		ϕ=60°

Table 2. Conditions of grinding test

Fig.20 shows the SEM picture of the ground workpieces. Obviously, the cylindrical workpiece having an original diameter of 0.6mm slimed down to one having dimension of 60μm in diameter, the aspect ratio of which is over 250. This result demonstrated that the constructed apparatus performed well even in actual micro-scale machining, and that micro-scale fabrication by ultrasonic-shoe centerless grinding technique is feasible.

Figure 20. SEM image of the ground tungsten carbide steel workpiece

6. Conclusions

Using ultrasonic vibration shoe centerless grinding technique to fabricate microscale tungsten carbide steel cylindrical workpiece was investigated. An experimental apparatus was modified by installing fine feed and adjustment mechanisms, and the ultrasonic vibration shoe used in experimental grinding was designed and made especially. The tungsten carbide steel cylindrical workpiece with 0.6mm in the original diameter and 15mm in the length was ground with a diamond grinding wheel. As a result, a microscale tungsten carbide steel cylindrical workpiece of around 60μm in diameter and 15mm in length, the aspect ratio of which was over 250. The validity of the new microscale tungsten carbide steel cylindrical workpiece using ultrasonic vibration shoe centerless grinding technique is confirmed.

For a new microscale tungsten carbide steel cylindrical workpiece technology, it is further required to investigate influencing factors, such as the workpiece geometrical arrangement, and ultrasonic vibration amplitude how to affect the machining accuracy, i.e., workpiece roundness. The more fine adjustment mechanism and the measurement rig suitable for micro-scale workpiece less than 100◎m also be designed and made.

Future work will focus on further developing this new technique in terms of the influence of grinding conditions, such as the workpiece geometrical configuration(ϕ, γ), on machining accuracy,i.e., workpiece roundness. To this end, it is essential to first develop a roundness measurement method for microscale components less than100μm in diameter since there is, at present, no commercially available measurement rig suitable for such microscale components. This work will be detailed in a future report.

Author details

Yufeng Fan
School of Mechanical & Automotive Engineering Zhejiang, University of Science and Technology, Hangzhou, China

7. References

[1] Maruo S, Ikuta K. Two-photon micro stereolithography with submicron resolution-fabrication of a freely movable mechanism. In: Eighth international conference on rapid prototyping. 2000. p. 201.
[2] Yamaguchi K, Nakamoto T, Abraha P, Karyawan, Ito A. Manufacturing of micro-structure using ultraviolet ray photoactive resin (3rdReport, Beam shape and hardening characteristics in focused beam drawing method). Trans Jpn Soc Mech Eng Ser C 1995;61(581):304.
[3] Nakai T, Marutani Y. Fabrication of resin model using ultraviolet laser. J Jpn Soc Technol Plast 1988; 29(335):1249.
[4] Vasile MJ, Grigg DA, Griffith JE, Fizgerald EA, Russell PE. Scanning probe tips formed by focused ion beams. Rev Sci Instrum 1991; 62(9):2167.
[5] Heeren P-H, Reynaerts D, Van Brussel H, Beuret C, Larsson O, Bertholds A. Microstructuring of silicon by electro-discharge machining (EDM)—part II. Appl Sens Actuators 1997;A61:379.
[6] Masuzawa T, Fujino M, Kobayashi K. Wire electro-discharge grinding for micro machining. Ann CIRP 1985;34(1):431.
[7] Masuzawa T, Fujino M. A process for manufacturing very fine pin tools. SME Technical Paper, MS90;1990. p.307/1–11.
[8] Uehara Y, etal. Development of small tool by micro fabrication system applying ELID grinding technique. In: Initiatives of precision engineering at the beginning of millennium, JSPE2001:10.
[9] Okano K, WaidaT, Suto T, Mizuno J, Kobayashi T. Micro-grinding of micromachine parts. In: Proceedings of international conference on abrasive technology. 1993. p.100.
[10] Yamagata Y, Higuchi T. Three-dimensional micro fabrication by precision cutting technique. J JSPE 1995; 61(10):1361.
[11] Hashimoto F. Effects of friction and wear characteristics of regulating wheel on centerless grinding. SME Technical Papers, MR99-226; 1999. p.1–10.
[12] Wu Y, Fan Y, Kato M, Wang J, Syoji K, Kuriyagawa T. A new centerless grinding technique without employing a regulating wheel. Key Eng Mater 2003; 238–239: 355–360.
[13] Fan Y, Wu Y, Kato M, Tachibana T, Syoji K, Kuriyagawa T. Design of an ultrasonic elliptic-vibration shoe and its performance in ultrasonic elliptic-vibration-shoe centerless grinding. JSME Int J Ser C 2004;47(1):43–51.

[14] Wu Y, Fan Y, Kato M, Kuriyagawa T, Syoji K, Tachibana T. Determination of an optimum geometrical arrangement of workpiece in the ultrasonic elliptic-vibration shoe centerless grinding. Key Eng Mater 2004; 257–258:495–500.
[15] Wu Y, Fan Y, Kato M, Kuriyagawa T, Syoji K, Tachibana T. Development of an ultrasonic elliptic vibration shoe centerless grinding technique. J Mater Process Technol 2004;155–156:1780–7.
[16] Ueha S, Tomikawa Y. New ultrasonic motor. Tokyo: Sougoudenshi Publications; 1991 [in Japanese].
[17] Kenjo N, Yubita S. Primer of ultrasonic motor. Tokyo: Sougoudenshi Publications; 1991[in Japanese].
[18] Piezoelectric ceramics technical handbook. Tokyo: Fuji Ceramics;1998[in Japanese].